urn on or before the
ate stamped below

KINGSTON COLLEGE

00080009

GIS and Health

Also in the GISDATA Series

I   **GIS and generalization: methodology and practice**
    edited by J. C. Müller, J.-P. Lagrange and R. Weibel

II **Geographic objects with indeterminate boundaries**
    edited by P. A. Burrough and A. U. Frank

III **GIS diffusion: the adoption and use of geographical information systems in local government in Europe**
    edited by I. Masser, H. Campbell and M. Craglia

IV **Spatial analytical perspective on GIS**
    edited by M. Fischer, H. J. Scholten and D. Unwin

V   **European geographic information infrastructures: opportunities and pitfalls**
    edited by P. Burrough and I. Masser

Series Editors

I. Masser and F. Salgé

GIS and Health

EDITORS

**ANTHONY C. GATRELL
AND
MARKKU LÖYTÖNEN**

GISDATA VI

SERIES EDITORS

I. MASSER and F. SALGÉ

UK   Taylor & Francis Ltd, 1 Gunpowder Square, London, EC4A 3DE
USA   Taylor & Francis Inc., 325 Chestnut Street, 8th Floor, Philadelphia, PA 19106

Copyright © Taylor & Francis 1998

All rights reserved. No part of this publication may be reproduced, stored in a retrieval system, or transmitted in any form or by any means, electronic, electrostatic, magnetic tape, mechanical, photocopying, recording or otherwise, without the prior permission of the copyright owner.

British Library Cataloguing in Publication Data

A catalogue record for this book is available from the British Library.
ISBN 0-7484-07790 (cased)

Library of Congress Cataloguing-in-Publication Data are available

Cover design by Hybert Design and Type, Waltham St Lawrence, Berkshire.

Typeset in Times 10/12pt by Santype International Ltd, Salisbury, Wiltshire.

Printed and bound by T.J. International Ltd, Padstow, UK

Contents

Series Editors' Preface	vii
Editors' Preface	ix
Notes on Editors	xi
Contributors	xii
European Science Foundation	xv
PART ONE **Methodological Issues**	1
1 **GIS and Health Research: An Introduction** Anthony C. Gatrell and Markku Löytönen	3
2 **GIS as an Enabling Technology** Geoffrey M. Jacquez	17
3 **Spatial Statistics and the Analysis of Health Data** Robert Haining	29
4 **Statistical Methods for Spatial Epidemiology: Tests for Randomness** Martin Kulldorff	49
5 **Improving the Geographic Basis of Health Surveillance using GIS** Gerard Rushton	63
6 **Modelling Spatial Variations in Air Quality using GIS** Susan Collins	81
7 **GIS, Time Geography and Health** Markku Löytönen	97

PART TWO **Applications** 111

8 **GIS Applications for Environment and Health in Italy** 113
   Stefania Trinca

9 **A Multipurpose, Interactive Mortality Atlas of Italy** 125
   Mario Braga, Cesare Cislaghi, Giorgio Luppi and Carola Tasco

10 **Bayesian Analysis of Emerging Neoplasms in Spain** 139
   Gonzalo López-Abente

11 **The Development of an Epidemiological Spatial Information
   System in the Region of Western Pomerania, Germany** 153
   Nanja van den Berg

12 **Problems and Possibilities in the Use of Cancer Data by GIS – Experience
   in Finland** 167
   Lyly Teppo

13 **GIS in Public Health** 179
   Paul Wilkinson, Christopher Grundy, Megan Landon and Simon Stevenson

14 **Improving Health Needs Assessment using Patient Register Information in a
   GIS** 191
   Andrew Lovett, Robin Haynes, Graham Bentham, Sue Gale, Julii Brainard
   and Gisela Suennenberg

15 **Conclusions** 205
   Anthony C. Gatrell and Markku Löytönen

   Index 209

The GIS Data Series

Series Editors' Preface

Over the last few years there have been many signs that a European GIS community is coming into existence. This is particularly evident in the launch of the first of the European GIS (EGIS) conferences in Amsterdam in April 1990, the publication of the first issue of a GIS journal devoted to European issues (*GIS Europe*) in February 1992, the creation of a multipurpose European ground-related information network (MEGRIN) in June 1993, and the establishment of a European organisation for geographic information (EUROGI) in October 1993. Set in the context of increasing pressures towards greater European integration, these developments can be seen as a clear indication of the need to exploit the potential of a technology that can transcend national boundaries to deal with a wide range of social and environmental problems that are also increasingly seen as transcending the national boundaries within Europe.

The GISDATA scientific programme is very much part of such developments. Its origins go back to January 1991, when the European Science Foundation funded a small workshop at Davos in Switzerland to explore the need for a European level GIS research programme. Given the tendencies noted above it is not surprising that participants of this workshop felt very strongly that a programme of this kind was urgently needed to overcome the fragmentation of existing research efforts within Europe. They also argued that such a programme should concentrate on fundamental research and it should have a strong technology transfer component to facilitate the exchange of ideas and experience at a crucial stage in the development of an important new research field. Following this meeting a small coordinating group was set up to prepare more detailed proposals for a GIS scientific programme during 1992. A central element of these proposals was a research agenda of priority issues groups together under the headings of geographic databases, geographic data integration, and social and environmental applications.

The GISDATA scientific programme was launched in January 1993. It is a four-year scientific programme of the Standing Committee of Social Sciences of the European Science Foundation. By the end of the programme more than 300 scientists from 20 European countries will have directly participated in GISDATA activities and many others will have utilised the networks built up as a result of them. Its objectives are:

- to enhance existing national research efforts and promote collaborative ventures overcoming European-wide limitations in geographic data integration, database design and social and environmental applications;
- to increase awareness of the political, cultural, organisational, technical and informational barriers to the increased utilisation and inter-operability of GIS in Europe;
- to promote the ethical use of integrated information systems, including GIS, which handle socio-economic data by respecting the legal restrictions on data privacy at the national and European levels;
- to facilitate the development of appropriate methodologies for GIS research at the European level;
- to produce output of high scientific value;
- to build up a European network of researchers with particular emphasis on young researchers in the GIS field.

A key feature of the GISDATA programme is the series of specialist meetings that has been organised to discuss each of the issues outlined in the research agenda. The organisation of each of these meetings is in the hands of a small task force of leading European experts in the field. The aim of these meetings is to stimulate research networking at the European level on the issues involved and also to produce high quality output in the form of books, special issues of major journals and other materials.

With these considerations in mind, and in collaboration with Taylor & Francis, the GISDATA series has been established to provide a showcase for this work. It will present the products of selected specialist meetings in the form of edited volumes of specially commissioned studies. The basic objective of the GISDATA series is to make the findings of these meetings accessible to as wide an audience as possible to facilitate the development of the GIS field as a whole.

For these reasons the work described in the series is likely to be of considerable importance in the context of the growing European GIS community. However, given that GIS is essentially a global technology most of the issues discussed in these volumes have their counterparts in research in other parts of the world. In fact there is already a strong UK dimension to the GISDATA programme as a result of the collaborative links that have been established with the National Center for Geographic Information and Analysis through the United States National Science Foundation. As a result it is felt that the subject matter contained in these volumes will make a significant contribution to global debates on geographic information systems research.

Ian Masser
François Salgé

Editors' Preface

Among the many domains of application for GIS technology and methodology, few can be as significant as that of health. Whether helping in the detection of tendencies for diseases to show departures from non-randomness, the identification of areas of elevated relative risk, or associations between disease incidence and social and environmental factors, GIS has a role to play. GIS also has a role to perform in health care planning and the efficient and equitable distribution of resources, and while this parallel area of application is, deliberately, not emphasised here, it is nonetheless important and merits full attention elsewhere. The emphasis here, then, is very much on the relevance of GIS for public health, and both the technical contributions and the applications (broadly epidemiological) reflect this.

A small task force, comprising Professor Tony Gatrell (Lancaster University, UK), Professor Markku Löytönen (University of Turku, Finland), and Dr Max Craglia (University of Sheffield, UK) met in late 1995 to begin to organise a specialist meeting on GIS and health. A total of 16 participants were invited to the meeting, which was held near Helsinki, in June 1996. In keeping with the ethos of other GISDATA meetings, the main aim was to bring together a number of scholars, mostly from Europe but with some participation from the USA, who were either using GIS in their health-related research or who were using health as a platform or testing ground for their developments of new GIS-linked methodology. As is clear from the list of contributors, the research area of GIS and health is no respecter of arbitrary disciplinary boundaries, and the skills of geographers, statisticians, epidemiologists and environmental scientists are all brought to bear in what follows.

The work presented in this volume inevitably represents only a partial picture of the 'state-of-the-art' of GIS and health, since the remit of the programme was to ensure as wide as possible a representation from across Europe, with emphasis given to countries within the EU.

In seeking to put together a list of invitations, it became apparent that, despite the buoyancy of health-related GIS work in some countries of Europe, there was a relative dearth of such activity in others. For example, France, Portugal, Belgium and Denmark – to mention only a few – are not represented in this volume. We hope that publication will give encouragement to those working across Europe to consider pursuing this area of research and, where appropriate, to forge links with those European colleagues already working in this field.

Tony Gatrell
Markku Löytönen

Notes on Editors

Tony Gatrell graduated with a First Class Honours degree in Geography from the University of Bristol. At Bristol his interests in quantitative human geography developed, largely under the influence of Peter Haggett. He then spent four years at Pennsylvania State University, completing a Masters and a PhD under the supervision of Peter Gould. He returned to the UK to take up a Lectureship in Human Geography at Salford University before moving to Lancaster University in 1984. He is now Professor of the Geography of Health and is currently on secondment to direct the Institute for Health Research at Lancaster. His research interests are in the geography of health, focussing in particular on geographical epidemiology and health inequalities. The use of GIS in both areas is of special interest.

Markku Löytönen graduated from the University of Helsinki with a degree in Human Geography. He remained at Helsinki to complete his PhD on the spatial development of intercommunications systems in Finland, 1860–1980. He became Assistant Professor, and later docent (senior lecturer) at the University of Helsinki. He is currently Associate Professor of Geography at the University of Turku, maintaining his docentship in Helsinki. His research interests range from the history of geography and exploration to quantitative methods and GIS with special interest in the geography of health. He uses GIS in his research on forecasting epidemics and analysing longitudinal data.

Contributors

Graham Bentham
School of Environmental Sciences, University of East Anglia, Norwich NR4 7TJ, UK

Mario Braga
Instituto Dermopatico del l'Immacolata, viai dei Monti di Creta 104, 00167 Roma, Italy

Julii Brainard
School of Environmental Sciences, University of East Anglia, Norwich NR4 7TJ, UK

Cesare Cislaghi
Instituto di Biometria e Statistica Medica, via Venezian 1, 20121 Milano, Italy

Susan Collins
Department of Geography and Sheffield Centre for Geographic Information and Spatial Analysis, University of Sheffield, Sheffield S10 2TN, UK

Sue Gale
School of Environmental Sciences, University of East Anglia, Norwich NR4 7TJ, UK

Anthony C. Gatrell
Institute for Health Research, Lancaster University, Lancaster LA1 4YT, UK

Christopher Grundy
Environmental Epidemiology Unit, Department of Public Health and Policy, London School of Hygiene and Tropical Medicine, Keppel Street, London WC1E 7HT, UK

Robert Haining
Department of Geography and Sheffield Centre for Geographic Information and Spatial Analysis, University of Sheffield, Sheffield S10 2TN, UK

Robin Haynes
School of Environmental Sciences, University of East Anglia, Norwich NR4 7TJ, UK

Geoffrey M. Jacquez
BioMedware, 516 North State Street, Ann Arbor, Michigan 48104, USA

Martin Kulldorff
Biometry Branch, National Cancer Institute, 6130 Executive Boulevard, Bethesda, Maryland 20892, USA

Megan Landon
Environmental Epidemiology Unit, Department of Public Health and Policy, London School of Hygiene and Tropical Medicine, Keppel Street, London WC1E 7HT, UK

Gonzalo López-Abente
Centro Nacional de Epidemiologia, Instituto de Salud Carlos III, Sinesio Delgado 6-28029, Madrid, Spain

Andrew Lovett
School of Environmental Sciences, University of East Anglia, Norwich NR4 7TJ, UK

Markku Löytönen
Department of Geography, University of Turku, FIN-20014 Turku, Finland

Giorgio Luppi
Ufficio Statistica – Regione Emilia-Romagna, Viale Silvani 4/3, Bologna, Italy,

Gerard Rushton
Department of Geography, University of Iowa, Iowa City, Iowa 52242, USA

Simon Stevenson
Environmental Epidemiology Unit, Department of Public Health and Policy, London School of Hygiene and Tropical Medicine, Keppel Street, London WC1E 7HT, UK

Gisela Suennenberg
School of Environmental Sciences, University of East Anglia, Norwich NR4 7TJ, UK

Carola Tasco
Osservatorio Epidemiologico – Regione Lombardia, via Stresa 3, 20100 Milano, Italy,

Lyly Teppo
Finnish Cancer Registry, Liisankatu 21B, FIN-00170, Helsinki, Finland

Stefania Trinca
Servizio Elaborazione Dati, Istituto Superiore di Sanità, Viale Regina Elena 299, 00161 Roma, Italy

Nanja van den Berg
Zentrum für Raumbezogene Informationsverarbeitung, Brandteichstrasse 19, D17489, Greifswald, Germany

Paul Wilkinson
Environmental Epidemiology Unit, Department of Public Health and Policy, London School of Hygiene and Tropical Medicine, Keppel Street, London WC1E 7HT, UK

The **European Science Foundation** is an association of its 55 member research councils, academies, and institutions devoted to basic scientific research in 20 countries. The ESF assists its Member Organisations in two main ways: by bringing scientists together in its Scientific Programmes, Networks and European Research Conferences, to work on topics of common concern: and through the joint study of issues of strategic importance in European science policy.

The scientific work sponsored by ESF includes basic research in the natural and technical sciences, the medical and biosciences, and the humanities and social sciences.

The ESF maintains close relations with other scientific institutions within and outside Europe. By its activities, ESF adds value by cooperation and coordination across national frontiers and endeavours, offers expert scientific advice on strategic issues, and provides the European forum for fundamental science.

This volume is the sixth in a series arising from the work of the ESF Scientific Programme on Geographic Information Systems: Data Integration and Database Design (GISDATA). The programme was launched in January 1993 and through its activities has stimulated a number of successful collaborations among GIS researchers across Europe.

Further information on the ESF activities in general can be obtained from:
European Science Foundation
1 quai Lezay Marnesia
67080 Strasbourg Cedex
tel: +33 88 76 71 00
fax: +33 88 37 05 32

EUROPEAN SCIENCE FOUNDATION

This series arises from the work of the ESF Scientific Programme on Geographic Information Systems: Data Integration and Database Design (GISDATA). The Scientific Steering Committee of GISDATA includes:

Dr Antonio Morais Arnaud
Faculdade de Ciencas e Tecnologia
Universidade Nova de Lisboa
Quinta da Torre, P-2825 Monte de Caparica
Portugal

Professor Hans Peter Bähr
Universität Karlsruhe (TH)
Institut für Photogrammetrie und Fernerkundung
Englerstrasse 7, Postfach 69 80
(W) 7500 Karlsruhe 1
Germany

Professor Kurt Brassel
Department of Geography
Universitity of Zurich
Winterthurerstrasse 190
8057 Zurich
Switzerland

Dr Massimo Craglia (Research Coordinator)
Department of Town & Regional Planning
University of Sheffield
Western Bank, Sheffield S10 2TN
UK

Professor Jean-Paul Donnay
Université de Liège, Labo. Surfaces
7 place du XX août (B.A1-12)
4000 Liège
Belgium

Professor Manfred Fischer
Department of Economic and Social Geography
Vienna University of Economic and Business Administration
Augasse 2-6, A-1090 Vienna, Austria

Professor Michael F. Goodchild
National Center for Geographic Information and Analysis (NCGIA)
University of California
Santa Barbara, California 93106
USA

Professor Einar Holm
Geographical Institution
University of Umeå
S-901 87 Umeå
Sweden

Professor Ian Masser (Co-Director and Chairman)
Department of Town & Regional Planning
University of Sheffield
Western Bank, Sheffield S10 2TN
UK

Dr Paolo Mogorovich
CNUCE/CNR
Via S. Maria 36
50126 Pisa
Italy

Professor Nicos Polydorides
National Documentation Centre, NHRF
48 Vassileos Constantinou Ave.
Athens 116 35
Greece

M. François Salgé (Co-Director)
IGN
2 ave. Pasteur, BP 68
94160 Staint Mandé
France

Professor Henk J. Scholten
Department of Regional Economics
Free University
De Boelelaan 1105
1081 HV Amsterdam
Netherlands

Dr John Smith
European Science Foundation
1 quai Lezay Marnesia
67080 Strasbourg
France

Professor Esben Munk Sorensen
Department of Development and Planning
Aalborg University, Fibigerstraede 11
9220 Aalborg
Denmark

Dr Geir-Harald Strand
Norwegian Institute of Land Inventory
Box 115, N-1430 Ås
Norway

Dr Antonio Susanna
ENEA DISP-ARA
Via Vitaliano Brancati 48
00144 Roma
Italy

PART ONE

Methodological Issues

CHAPTER ONE

GIS and Health Research: An Introduction

ANTHONY C. GATRELL AND MARKKU LÖYTÖNEN

1.1 Introduction

There is a long and rich tradition of investigating the spatial patterning of health events and disease outcomes, stretching back at least as far as the nineteenth century. Such investigations have used classical tools of visualisation, as well as methods of data exploration and modelling drawn from the statistical and epidemiological sciences. Paralleling such studies in geographical and environmental epidemiology have been others, set less within a natural science paradigm and more within a social science context, that have involved a study of health variations and inequalities, not only in terms of health outcomes but also in terms of access to, and the provision of, services. What both areas of research have in common is the recognition that space and place 'make a difference'.

Much of this work predates the revolution in spatial data-handling that has come about because of the development of Geographical Information Systems (GIS). The question then arises as to what extent health research can be given added value by using the tools of modern GIS. Here, we aim to review some of the ways in which GIS and health research come together, but emphasising the research needs at the end of the century, in a European setting. This is not the first time that GIS and health research have been discussed in a European context. For example, the Regional Office of WHO convened a meeting of health and GIS professionals in 1990, at the National Institute of Public Health and Environmental Protection in Bilthoven. One outcome was to suggest the setting up of a Health and Environment GIS (HEGIS), the nature of which is discussed in de Lepper *et al.* (1995, pp. 333–348).

We need to be clear about precisely what GIS involves, since there still seems to be a perception among some that it involves simply desktop mapping systems. We do not seek to denigrate the important role of visualisation (indeed, we give it due acknowledgment later); it plays a valuable role in hypothesis generation, for example. But a comprehensive definition is that GIS are systems for the collection, storage, integration, analysis and display of spatially-referenced data; in the present context such data are those representing, or associated with, health and disease.

Below, we touch on data collection, say nothing about data storage, say something about integration, quite a lot about analysis (historically, rather neglected in GIS) and something about display or visualisation.

Implicit in our opening remarks is a tension in health research between those engaged in looking at disease incidence from a biomedical standpoint and those approaching their studies from the viewpoint of health as a social product. Put crudely, are we studying the geography of disease – and thereby drawing on a biomedical model of enquiry – or are we engaged in a study of the geography of health, where as much priority is given to lay perceptions of health and illness as to quantitative expressions of ill-health? It is important to recognise that many of the widely-used health indicators, such as measures of standardised mortality, paint a very partial picture of either disease incidence or health status. Further, we tend to dwell usually on areas of high disease risk, often neglecting to note that areas with low rates may also give clues about disease causation. Morbidity data will often be more valuable, though these too may be subject to variations in diagnosis behaviour, and there may be intra-national as well as inter-national variations in the quality and availability of such data. But there may be scope for collecting data on qualitative health variations and perceptions, and the way in which these might be incorporated in a GIS framework bears examination. There might also be greater scope for using GIS to paint a picture of 'health' as well as ill-health, though given that health is a 'contested concept' this may be a vain hope!

That said, GIS has most commonly been used within a biomedical framework, in studies of disease incidence. But what sorts of disease should we focus effort on? Those which attract most publicity are often those whose incidence is actually rather low. For example, in Britain there have been press reports in the past three years about children born with eye malformations and with limb defects, and, very recently, about meningitis outbreaks. Also, enormous research effort has gone into the study of leukaemia in children, a research area that produced the first really original investigations into the use of GIS in a health context (Openshaw et al., 1987). However, devastating as these diseases are, in terms of numbers of people affected they pale into insignificance compared with the incidence of lung and breast cancer, or mental health problems, for example.

What, then, is covered in this edited collection, and what is ignored? Nothing is said here about the use of GIS in the study of communicable disease or links to forecasting models (see, for example, Cliff et al., 1986) and GIS; one could argue that this work has developed well in the absence of 'GIS input', so the 'added value' (de Lepper et al., 1995) provided by GIS is perhaps very marginal. Nor do we say much about service provision, or using GIS as a decision support tool to plan the spatial configuration of services. Also omitted from the chapters is any focus on the provision of emergency services, or planning responses by health service workers to major disasters and emergencies. Some of these applications are considered in another overview of the field (Gatrell and Senior, 1998).

A major aim of the GISDATA research programme is the wish to bring together experts from many different backgrounds and sciences. A study of geographical variations in health across Europe requires the skills and expertise of geographers, environmental scientists, statisticians, epidemiologists and sociologists, to name but a few. No one discipline can speak with authority in such a diverse and socially relevant field. Having begun to establish the beginnings of a wider European network of those with a common research interest, the next aim is to focus on a set of research themes and to identify some priority areas for research at the end of this century and the start of the next. This introduction offers a few initial thoughts, but the hope is that like-minded individuals begin to collaborate, to exchange experience and to provide cross-national input into major European initiatives.

One such initiative is already in the pipeline. This is the ESF initiative on Environment and Health (ESF, 1995), which is addressing a broad set of research issues concerned with the interaction between environmental quality and human health. A question we need to address is to what extent GIS can 'feed into' this work? To some extent these issues have already been debated (de Lepper et al., 1995), but this new initiative calls for further inputs. And, as the authors themselves acknowledge (ESF, 1995) the 'environment' is a social setting as well as a physical one, with socioeconomic and lifestyle factors playing as much of a role in disease incidence as environmental pollution. For example, if we wish to understand the observed pattern of throat cancer we will need information not only on exposure to dioxins but also on cigarette and alcohol consumption.

We structure the present chapter into two main sections: one of these is concerned with environmental and geographical epidemiology. The second is more concerned with a social science perspective on health, where the focus is on health inequalities and health 'promotion'. We conclude with our own tentative indication of where we see promising areas of research.

1.2 GIS in environmental and geographical epidemiology

One major area in which GIS and health research have come together is via the study of environmental epidemiology. Here, we look for links between disease and the physical environment, while controlling for the impact of lifestyle factors such as smoking, diet and physical exercise. We can contrast this with the more narrowly defined geographical or spatial epidemiology, where description, exploration and modelling of disease incidence does not necessarily involve making direct links to environmental contamination. Here, studies of disease clustering, of cluster identification, of association with potential point and line sources of pollution, and of space–time disease incidence, are given priority (see Elliott et al., 1992, for numerous examples). Ideally, these should use data at a fine level of spatial resolution and should relate to the individuals themselves (disease 'cases', possibly compared with 'controls').

1.2.1 Environmental epidemiology

There is a considerable literature on using GIS to explore environmental correlates of disease. ESF (1995) has called in particular for further work on risks from ambient and indoor air pollutants; those due to vehicular traffic and radon are mentioned explicitly; and on reproductive toxicology (birth defects, environmental oestrogens). There is clearly scope for a GIS input here, especially in the first area. What, for example, do we know about low-level ozone across Europe? In the UK the Photochemical Oxidants Research Group has used data from a network of monitoring sites to create interpolated maps of risk. What is the current configuration of monitoring sites in other European countries? To what extent is it possible, in 1998, to assess broad-scale correlations between, say, the incidence of asthma and ozone levels across the continent? Where are the standard errors of prediction of ozone levels particularly high, suggesting the need for additional monitoring sites?

Some work on using GIS to explore links between air pollution and health has been reported by Dunn et al., (1995), and also by Kingham (1993), who has sought to link models of air pollution to GIS, using output from the models to define areas of exposure and then relating disease incidence to these rather than simply drawing circular buffer zones around a possible point source and assuming this defines an appropriate area of risk. This work needs building on. We see one of the primary areas for research that of integrating environmental modelling and health databases, within a GIS framework.

Within this environmental epidemiological framework we also need to give due weight to human behaviour. People do not sit at home waiting to be polluted by air! This social dimension also arises in other contexts, such as food poisoning (mainly salmonella and campylobacter), an area of investigation also flagged by ESF (1995). Some work has been done on this in the UK from a GIS perspective (Brown et al., 1995), but more needs to be done on the relative importance of behavioural factors (such as food preparation in the home) and environmental factors (contamination of water supplies). ESF have also called for work on deaths and injuries from all forms of accidents, and on urban health. While some work in such areas falls very much within environmental epidemiology (for example, links between air pollution and respiratory disease in large cities), in both areas we need data both on the socioeconomic backcloth and on individuals in order to assess the relative importance of environmental and social variables. Put crudely, is the incidence of asthma in large European cities a function of exposure to air pollutants or to exposure to poor housing conditions, for example? We return to this theme later.

1.2.2 Geographical epidemiology

Given a set of individuals diagnosed with some disease, together with some form of address reference, we can aggregate the cases into any fixed set of areal units (counties, departments, communes, or whatever) and define measures of relative risk, such as standardised mortality or morbidity ratios, given appropriate denominators. One virtue of GIS is that we are freed from the tyranny of census-based areal units; we might explore disease risk around a point source of pollution, or along a busy motorway instead. However, this does lead to severe problems of data integration or 'areal interpolation', reviewed by Flowerdew and Green (1991), for example. Once we have done this we can bring to bear various mapping and smoothing techniques (such as empirical Bayes estimation: see Langford, 1994; Bailey and Gatrell, 1995) to help interpret the data. Tests of spatial (auto)correlation are also likely to prove useful. Such tools (see Douven and Scholten, 1995, for a review) can be useful at a variety of scales, varying from broad-scale investigations across Europe (see, for example, the European Cancer Atlas) to national cancer atlases (including several very distinguished examples), down to small-scale studies within local areas.

However, if we have addresses of individual cases (and, ideally, matching controls) it seems perverse, from an analytical point of view, to lose this detail. As a result, several groups of researchers (for example, Rushton in the USA, Kulldorff in Sweden, and Lawson in the UK; see Bailey and Gatrell, 1995, and Gatrell *et al.*, 1996, for a partial review) have been exploring the use of modern point pattern methods in exploring and modelling disease risk. This has included the use of so-called second-order methods for detecting whether there is disease clustering; the tendency for cases to cluster or aggregate more than the population at risk. Extensions to this have examined space–time clustering; that is, whether events that are spatially proximate are also 'close' in time, a methodology that has been applied to the study of many diseases, for example Legionnaires' disease (Bhopal *et al.*, 1992). A different problem in geographical epidemiology is that of detecting clusters. Here, the classic work is by Openshaw and co-workers (1987), but other approaches include the use of density or 'kernel' estimation to define a continuous surface of disease risk and relative risk; see Kelsall and Diggle (1995) for a modern view, and Bailey and Gatrell (1995) for an introduction to this method. Yet another problem is that of assessing whether there is a raised relative risk around some pre-specified point or line source of potential pollution. Again, there are various statistical methodologies proposed in the literature; see for example Bithell and Stone (1989), Diggle *et al.*, (1990) and Diggle and Rowlingson (1994).

What are the data requirements for such investigations? Generally, such work will take place at the small area level, and the need for comparative data across more than one country may be slight. But some of the methods, for example kernel estimation, could be employed cross-nationally, for example looking at the incidence of neural tube defects, or road traffic accidents, or asthma on the French–Belgian–German border. Quite apart from national differences in diagnosis, to what extent would this work be hampered by variations in spatial referencing? At what level of resolution are such data available? In the UK the unit postcode can be matched to grid references of 100 or 10 metres; in some instances (and subject to confidentiality restrictions) data may be available down to 1 m resolution. The OS ADDRESS-POINT product permits this if the full address of a patient is known. We do not want to dwell over-long on issues of data integration across national boundaries, nor do we wish to say too much about issues of data quality, since these have been addressed in other GISDATA meetings, but the issue does need raising here. So too do issues of data availability. As pressures grow for legislation on data protection, what are the prospects for obtaining access to (suitably anonymised) patient records? It may be easier to get access to environmental data for some studies, though such data bring with them their own problems of 'fitness for purpose', and the problems of data quality (for example, in measuring dioxins in soils) are well known.

One of the key issues that arises in geographical epidemiology concerns the representation of exposure. Typically, this is represented by address at date of diagnosis. But this may be a far from perfect representation. How can GIS help us to represent exposures in other contexts (such as the workplace)? Can we model action and activity spaces, for example within a time–space framework suggested originally by Torsten Hägerstrand? Can we incorporate information about residential histories into our analyses? Perhaps we can use measures of exposure such as years spent at a particular location rather than current address, address at diagnosis, or address at birth. Research into the geographical patterning of motor neurone disease, based at Lancaster University, is attempting to do just this (Sabel and Gatrell, 1998). It parallels some fascinating work on multiple sclerosis in Norway (Riise et al., 1991) and also by David Barker in Southampton (Barker, 1994), the work of whose team suggests that we can only begin to understand the contemporary pattern of, for instance, cardiovascular disease, if we study the individual's earlier social environment.

1.2.3 Software tools

What software tools are currently available to help the kinds of analysis we seek to do? Here, we need to distinguish between software environments for spatial data analysis and those for GIS. A case can be made for arguing that the former are required for geographical epidemiology, while GIS 'comes into its own' in an environmental epidemiological context. Spatial analysis of epidemiological data has proceeded quite comfortably without GIS! The added value of GIS is in the linking of databases; for example, in seeking to explore links between cancers and high voltage power lines, where different spatial databases are required. But if we seek solely to establish whether there is clustering of disease, or whether there is raised incidence of cancer around an incinerator (Diggle et al., 1990) we can do such analysis using software for interactive spatial data analysis. Such software includes that for spatial point pattern analysis (SPLANCS, run within S-Plus: see Rowlingson and Diggle, 1993) and, more generally, packages and environments such as INFO-MAP (Bailey and Gatrell, 1995) and LISP-STAT (Tierney, 1990). If we particularly wish to employ a GIS then there are links between some proprietary GIS and statistical software (for example, S-Plus for ARC/INFO).

Other tools for visualisation are required that free us from an over-reliance on conventional choropleth maps. Good examples are cartograms or 'isodemographic maps', which have been used for many years by epidemiologists (Selvin et al., 1988; Dorling, 1995), and which express the area of a zone on a map in proportion to some measure of population at risk rather than physical area; as a result, large rural areas 'shrink' while more prominence is given to otherwise small urban areas in which most of the health events of interest are concentrated.

1.3 GIS and the 'new' public health

In Britain, Directors of Public Health are sometimes very much engaged in studies of a geographical or environmental epidemiological nature (as when, for example, there are local concerns expressed about 'cancer clusters'). However, the bulk of their work tends very much to be community-orientated and, at least in the 1990s, draws as much strength from a tradition of social medicine as from bio-medicine (Ashton, 1995). As suggested earlier, this 'new' public health sees health as more than the avoidance of early death, convenient though it is to measure this! It gives due weight to quality of life for the individual in the community, recognising that this is partly to be explained by individual decision-making but also by the wider socioeconomic settings (at both a national and local scale) within which the individual is situated.

Much public health medicine, therefore, tackles issues at a fairly local level. In Britain health care planning and some provision takes place at a 'locality' scale and is, increasingly, primary care based. This means that there is an interest in health needs assessment at quite a small scale, requiring the use of census and other data in order to identify what services are required, and where. It also means that there is a growing interest in how well primary health care practitioners are delivering services.

GIS enters here in a number of different ways. Some use has been made of GIS in determining the boundaries of localities, for example (Bullen *et al.*, 1994). In other areas, census-based deprivation scores (for example, due to Townsend and Carstairs: see Morris and Carstairs, 1991) have been incorporated into health databases in order to identify areas of need. Such deprivation indices may also be used to paint often vivid pictures of health inequalities at small area level. What scope there might be for creating an index that permits comparison between European states is a topic worth further investigation. A substantial, and coherent programme of research into health 'variations' (inequalities) is currently underway in Britain, commissioned by the UK Economic and Social Research Council; there is a parallel initiative funded by the Department of Health. Some of the research projects within those programmes involve GIS.

What specific areas of public health might benefit from a GIS approach? Those which spring immediately to mind are the uptake of services for preventive medicine, such as childhood immunisation and the screening of breast and cervical cancers in women. In Britain these services are arranged via the general practitioner, though the 'delivery' of the care may well take place elsewhere. The question then arises to what extent does distance to the surgery or health care centre constrain the uptake of services? Such accessibility questions operate not just in remote rural areas (where use might be made of 'branch' or mobile services) but also in urban settings. These issues of accessibility suggest that GIS might prove useful as a framework within which to couch an investigation, since data on the road network and on patterns of public transport will be more sensitive than simply measuring straight-line distances. While some use has been made of notional travel speeds to define journey times along road networks, we are not aware of any work that uses timetables to define journey times by public transport.

Suppose we wish to examine the uptake of services by health centre, and relate this to the social environment or characteristics of that centre in order to see whether uptake is explained by neighbourhood deprivation. One of the difficult problems in doing so is that health centre or practitioner catchments, in Britain at least, do not follow census area boundaries. Rather, patients are drawn from different small areas; we will have census-based data for these, but unless we carry out expensive, large-scale surveys we will not have detailed individual data. Some work is being done on trying to circumvent this problem (Haynes *et al.*, 1995; Scrivener and Lloyd, 1995). The same issue arises in dental care too.

Others have argued that in order to understand health behaviour or outcomes we need to recognise influences at different scales. For example, respiratory illhealth may depend upon individual-level factors (such as smoking behaviour), on household factors (the presence of damp or mould), and on local environmental factors (such as traffic levels in adjacent streets). Some health researchers (for example, Jones and Duncan, 1995) have sought to make use of multi-level models as an appropriate modelling framework; one obvious research task is to embed these within a GIS framework, or at the very least to couple them to a proprietary GIS. Jones and Duncan (1995) have suggested that such models allow us to give due weight to the importance of 'place'. We need to recognise that the quality of local environments, and the services that are provided, has a very real impact on people's quality of life. Access to good, reasonably priced food, health and leisure facilities, crime-free zones, and so on (Sooman and Macintyre, 1995) are all important measures of one's well-being. We need to explore the potential of GIS to define areas, or preferably continuous surfaces, of differential access to health; not solely access to health services, important though this is.

1.4 Some research themes

In this final section we draw on what has been discussed earlier and relate some of these themes to the chapters that follow.

The methodological section begins with a critique by Geoff Jacquez of the scientific basis of much recent GIS-related research in health. He argues that too many applications have been technology-driven, and that the mere production of mapped output can reveal spatial patterns and associations that are quite spurious. Such patterns, he argues, need to be rigorously assessed using spatial statistical analyses, a task rendered difficult by the relative lack of such tools in proprietary systems and lack of an appreciation of the importance of such evaluation. Too many applications suffer, he claims, from what he calls a 'gee whiz' effect; data are mapped and apparently 'interesting' patterns lead to the drawing of possibly wholly inappropriate conclusions. Jacquez suggests that we could do a lot worse than revisiting some of Karl Popper's strictures on scientific explanation, setting up theories which one attempts constantly to refute or falsify. Whether or not one agrees with Jacquez that Popper's critical rationalist approach to scientific reasoning is the appropriate one to follow, it is surely the case that GIS researchers could profit from frequent critical reflection on their analyses and that the scientific basis of such research is given due consideration.

Jacquez' chapter sets the scene for the next three chapters, all of which in different ways address the need for more thorough and rigorous spatial analysis of health data. Bob Haining deals comprehensively with the analysis of health data that have been aggregated into a system of areal units. Invariably, such units comprise a 'patchwork quilt' of irregularly-shaped zones, the varying size and shape of which, together with their differing populations at risk, render spatial analysis a tricky task. Haining considers methods for both informal, exploratory data analyses, where pattern detection is required in order to answer the kinds of criticism raised by Jacquez, and also methods for fitting models to health data, particularly where there are covariates available that might explain spatial variation in health outcomes. In terms of exploratory data analysis the distinction between 'global' and 'local' measures is particularly important, since the use of a single number – what some call a 'whole-map' statistic – to represent the entire spatial distribution may mask a considerable amount of spatial heterogeneity.

But Haining also answers Jacquez' criticism of the lack of spatial analysis capability in GIS, since he reports on an important project (known as SAGE) that seeks to add spatial analysis functionality to a well-known proprietary GIS. Such functionality involves the ability to construct particular aggregations of areal units and to perform various exploratory data analyses; the links between tabular data, maps and other graphics (such as histograms and boxplots) facilitate the kind of interactive spatial data analysis that others (for example, Bailey and Gatrell, 1995) have promoted.

Coming from a background in medical and spatial statistics, Martin Kulldorff is less concerned with the overt links to GIS than with ensuring that the tests called for by Jacquez and others are carefully chosen and properly evaluated. His particular emphasis is on tests for spatial randomness, where it is important to distinguish between several problems, as noted above. Different methods are needed to ascertain whether there is disease clustering than if the detection of clusters is the focus of endeavour. Kulldorff reviews a number of such methods, both those for purely spatial analysis as well as extensions to the space-time domain. He does not consider explicitly the implementation of these methods within a GIS environment, but his work has implications for those who would wish to see such methods either embedded within proprietary systems or with some form of coupling to such systems.

Gerry Rushton follows Kulldorff in preferring an approach to spatial data that uses the geographical locations of the cases of illness or disease, rather than aggregating these to fixed systems of areal units. His goal is that of exploratory spatial analysis, conducted within a spatially continuous setting, but the audience Rushton has in mind is that of the public health specialist who has to evaluate 'cluster alarms' (a type of spatial problem also reviewed by Kulldorff). Rushton has developed and implemented his exploratory spatial analysis modules on CD-ROM, in order to assist such specialists in their work. One technique he develops is based on spatial filtering (what we have earlier called kernel estimation). This is illustrated using individual, address-matched data on infant mortality for the city of Des Moines in Iowa. The significance of particular 'clusters' is evaluated statistically, using Monte Carlo simulation. Such tools are likely to prove invaluable in disease surveillance systems.

There is a need, within environmental epidemiology, to link environmental monitoring and modelling (for instance, of air and water quality) to a GIS so that health event data can be associated with the modelled outputs. Susan Collins' work is motivated by this research need. But in order to establish links between air quality and health we need good spatial representations of the former; to achieve this by a dense network of monitoring stations is clearly expensive and in many cases unrealistic. Instead, we must rely on the outputs from running air dispersion models, but such models need evaluating against the available, but limited observable data. Collins examines two broad approaches to modelling air quality: one, a hybrid approach that links dispersion modelling with interpolation methods; the other, a regression-based approach that links GIS and statistical techniques. The interpolation (kriging) approach results in an over-smoothed map of pollution levels and a regression approach yields better results. The goal of a comprehensive software package, or at least a set of closely-coupled modules, that allows the user to run a dispersion model, define a possible area of exposure, associate this with health events, and fit some spatial statistical models to such data is surely not that far off.

As Collins' work has indicated, we seriously need, in geographical and environmental epidemiology, accurate models of exposure. We cannot always rely on residential address at the time of diagnosis if we wish to understand the processes giving rise to spatial patterning of disease events. We need to model activity patterns and use the results from these to inform epidemiological studies. This need is particularly acute in the case of possible links between vehicular traffic pollution and respiratory illness, the main motivation for Collins' work. We tend typically to assume fixed and certain locations of exposure; in reality these are fluid and uncertain. To what extent can we incorporate into our analyses recognition that exposure is 'fuzzy'; that there is a field of exposure rather than a fixed point represented by home address?

This kind of issue is considered by Markku Löytönen, who draws upon the rich vein of work on time–geography, originated by Torsten Hägerstrand and subsequently developed by his colleagues. The relevance of this is over both short and long time scales. In the short term, as indicated above, and especially in Section 1.2.2, we need adequate representations of the daily activity spaces of individuals, while over the long term we require detailed information on previous place of residence, and duration thereof. Löytönen provides an introduction to time geography for those to whom it is unfamiliar, and then considers recent and current progress being made in embedding time into GIS. Although only a hypothetical example, he goes on to explore the possibilities of linking data on migration to that on environmental radiation. Given the quite superb spatially-referenced data on historical migration, along with detailed measurements of radon gas levels and of caesium-137 fallout from Chernobyl, the time is ripe for some innovative work using GIS to examine, in a robust and meaningful way, links between environment and health.

Part 2 of the book considers the application of GIS to health problems in particular European settings. With two contributions from the UK, and two from Italy, coverage of applications across the European Union is inevitably selective and partial. Nonetheless, the chapters give an overview of the breadth of application of GIS-based analyses.

Stefania Trinca considers a number of such applications in Italy, drawn from geographical and environmental epidemiology, risk assessment, and public health in general. Her description of the Rome-based GEO.S.I.M system and, notably, its use of sophisticated Bayesian smoothing and modelling techniques, illustrates the calls made earlier by Haining for area-based analyses to be soundly based on best statistical practice. The fact that the system also permits analyses to detect possible elevated relative risks near point sources of pollution echoes Kulldorff's discussion of 'focused' tests of clustering. Trinca also draws attention to work on a system, EUPHIDS, for assessing the risk of exposure to pesticides, though, as she notes, there are difficulties in linking such exposures to data on morbidity and mortality; as with so many GIS-based applications, the incompatibility of areal units, and of geographical scales, renders analyses extremely problematic.

In a companion paper, Mario Braga develops further one of the applications areas considered by Trinca, that of visualising and exploring data on mortality. Braga's work is very much in the spirit of contemporary visualisation research in epidemiology, seeking to move away from static, paper-based mortality atlases, towards electronic, interactive products with spatial statistical capability. Comprising data for over 8000 municipalities in Italy, the interactive atlas Braga describes incorporates the ideas of kernel estimation outlined earlier and tests for spatial correlation in the data, and will shortly have the Bayesian estimation implemented. Braga illustrates the functionality of the atlas with reference to data on lung and stomach cancer in Tuscany and Lazio respectively.

The application of Bayesian estimation to epidemiological data is developed fully in the chapter by Gonzalo López-Abente, who applies these techniques to cancer mortality data in Spain. The particular cancers with which he is concerned are those of relatively low incidence, such as multiple myeloma, non-Hodgkin's lymphoma and connective tissue tumours, and López-Abente attempts to relate geographical variation in the incidence of these cancers to data on insecticides, herbicides, pesticides and other organic control agents. Since numbers of cases are small, even for the quite large provinces of Spain, Bayesian techniques are used, as in the Italian studies, to smooth the raw rates in order to avoid random fluctuations. But López-Abente goes beyond such exploratory methods to model his data using so-called Markov Chain–Monte Carlo (MCMC) methods. The model of relative risk allows for both unstructured and structured variation ('heterogeneity' and 'clustering' respectively), the latter allowing the relative risk in an area to be influenced by those in nearby zones. Output from the model includes both the estimated effects of the explanatory variables and an indication of whether there is significant heterogeneity and clustering.

López-Abente performs all his analyses without the aid of a proprietary GIS. However, like others keen to see more links between GIS and statistical spatial analysis he calls for both greater awareness of contemporary developments in such analysis and for the development of links between such tools and geographical information systems. But it should be noted that the data assembled by López-Abente for his study are based entirely on a fixed set of (quite large) areal units; the data are stored, conventionally, as flat files and the 'added value' of a GIS remains to be more fully explored. Nonetheless, the kind of analytical approach he espouses is 'state-of-the-art' in spatial analysis, and anyone using 'ecological' data on health needs to consider using such methods in their analyses.

Nanja van den Berg outlines the development of a project to create an epidemiological GIS for the region of Western Pomerania in Germany. The project is still in its early stages, concentrating on database development and preliminary visualisation of data (comprising information on asthma and other respiratory illnesses among children). Data are explored at a sub-regional (ZIP-code) scale, where van den Berg is faced with the common problem of the highly variable number of cases among the set of zones, and uses the chi-square statistic (cf. Brown *et al.*, 1995) as a transformation to map data on eczema, hay fever and asthma. Contrasts between urban and rural areas are highlighted. Data are also available at a finer scale, for the city of Greifswald, where individual addresses are obtained, and although no results are available at present there is clearly scope, as van den Berg indicates, for research to explore associations with attributes of cases and controls, as well as with environmental variables; it is here that the spatial information system will prove particularly valuable.

Lyly Teppo looks specifically at spatially-referenced data on cancer in Finland. Data from cancer registries are, in general, of very high quality, and since they deal with incidence rather than mortality, overcome a number of problems involving specifying cause of death. Even so, there are problems of variable diagnoses, coverage and accuracy to consider, and while such problems are unlikely to be serious from place to place within a country, variations between countries may well make analyses across borders somewhat problematic. Teppo also lays emphasis on the lengthy, and probably quite variable, latency periods for some cancers, echoing the calls made by Löytönen and others for due acknowledgment to be made of changing residential location. This is relatively unproblematic in Finland and other Nordic countries, and surely the number of GIS-based epidemiological studies in such countries will grow as researchers seek to exploit the wealth of good historical data available there. Teppo himself reviews a number of studies that exploit the spatial detail of the Finnish cancer data. These include: the possible associations between high voltage power lines and childhood cancer; possible links between exposure to fallout from Chernobyl and childhood leukaemia; and the relationship between exposure to indoor radon pollution and lung cancer.

Paul Wilkinson and his colleagues outline the research possibilities using highly disaggregated (postcode-based) data in Britain. Such data include not only those on mortality and cancer incidence, but also on births and congenital malformations, and on uptake of primary and community services. Denominator data, on the age–sex structure of populations, are available only for census units such as electoral wards or enumeration districts, and there are thorny problems in matching postcodes to census areas (see Scrivener and Lloyd, 1995, for example). Like others in the book, Wilkinson draws attention to the visualisation and analytical problems involved in examining small area data, reporting results from a ward-based analysis of hospital admissions for asthma. Despite the fact that such admissions are the 'tip of the iceberg' in asthma morbidity there are some suggestive correlations with socioeconomic measures, which Wilkinson suggests may be useful in needs assessment, a prime task in public health. Wilkinson also endorses points made elsewhere in the book, for example concerning 'cluster' detection (Jacquez, Kulldorff), surveillance of public health (Rushton), and the difficulties of adequate exposure assessment (Collins).

Andrew Lovett and his co-authors are also concerned with the nature and quality of census-based denominator data in Britain, but suggest that patient-based denominator data (from the National Health Service Central Register) may have a role to play in deriving useful health needs indicators. Ward-level resident populations at the time of the 1991 census are compared with NHSCR estimates on census day; the conclusion is that using patient registers to produce small area population estimates remains problematic, but that with improvements they should provide a valuable resource. Lovett also compares the use of census-based unemployment data with that on people claiming unemployment benefit, and the close match suggests that the latter may be a useful needs indicator, given that census information is only available every ten years. Finally, he reports the results of analyses that seek to derive census-based indicators for general practices (providing primary care at surgeries or clinics).

The chapters by Wilkinson and Lovett are an appropriate pair to bring the volume to a conclusion, since they begin to address issues concerned partly with environmental epidemiology but more with needs assessment. While we, like others, have made a distinction between applications of GIS that are primarily epidemiological, and those that are concerned with health care planning, it is clear that there is, in fact, a continuum of approaches, since identification of 'high risk' areas, for example, presumably demands fresh looks at resource allocation in order to address them. As a result, we perhaps need to concern ourselves in future work less with GIS and more with spatial *decision support* systems.

Acknowledgment

Jan Rigby, Department of Geography, Lancaster University, made some thoughtful comments on a first draft of the position paper for the specialist meeting, on which this introductory chapter is based.

References

ASHTON J. R. (1995) A vision of health for the North-West. Inaugural Lecture, University of Liverpool.
BAILEY T. C. and GATRELL, A. C. (1995) *Interactive Spatial Data Analysis*. Longman, Harlow.
BARKER D. J. P. (1994) *Mothers, Babies, and Disease in Later Life*. BMJ Publishing, London.
BHOPAL R., DIGGLE P. J. and ROWLINGSON B. S. (1992) Pinpointing clusters of apparently sporadic Legionnaires' disease. *British Medical Journal*, **304**, 1022–1027.
BITHELL J. F. and STONE R. A. (1989) On statistical methods for analysing the geographical distribution of cancer cases near nuclear installations. *Journal of Epidemiology and Community Health*, **43**, 79–85.
BROWN P. J. B., HIRSCHFIELD A. and MARSDEN J. (1995) Analysing spatial patterns of disease: some issues in the mapping of incidence data for relatively rare conditions, in de Lepper *et al.*, 1995, pages 145–163.
BULLEN N., MOON G. and JONES K. (1994) Defining communities: a GIS approach to delivering better health care. *Mapping Awareness*, **8**(2), 22–25.
CLIFF A. D., HAGGETT P. and ORD J. K. (1986) *Spatial Aspects of Influenza Epidemics*. Pion, London.
DE LEPPER M. J. C., SCHOLTEN H. J. and STERN R. M. (eds) (1995) *The Added Value of Geographical Information Systems in Public and Environmental Health*. Kluwer Academic Publishers, Dordrecht.
DIGGLE P. J., GATRELL A. C. and LOVETT A. A. (1990) Modelling the prevalence of cancer of the larynx in part of Lancashire: a new methodology for spatial epidemiology. Pages 29–45 in Thomas R. W. (ed.) *Spatial Epidemiology*, Pion, London.
DIGGLE P. J. and ROWLINGSON B. S. (1994) A conditional approach to point process modelling of elevated risk. *Journal of the Royal Statistical Society* Series A, **157**, Part 3, 433–440.
DORLING D. (1995) *A New Social Atlas of Britain*. John Wiley, London.
DOUVEN W. and SCHOLTEN H. J. (1995) Spatial analysis in health research, in de Lepper *et al.*, 1995, pages 117–133.
DUNN C. E., WOODHOUSE J., BHOPAL R. S. and ACQUILLA S. D. (1995) Asthma and factory emissions in northern England: addressing public concern by combining geographical and epidemiological methods. *Journal of Epidemiology and Community Health*, **49**, 395–400.
ELLIOTT P. J., CUZICK J., ENGLISH D. and STERN R. (1992) (eds) *Geographical and Environmental Epidemiology: Methods for Small Area Studies*. Oxford University Press, Oxford.
ESF (1995) Environment and Health Research Needs in Europe. Chairman's Report of a meeting held in association with WHO/EUROPE, Leicester, 24–26 May.
FLOWERDEW R. and GREEN M. (1991) Data integration: methods for transferring data between zonal systems. Pages 38–54 in Masser I. and Blakemore M. (eds) *Handling Geographical Information: Methodology and Potential Applications*, Longman, Harlow.
GATRELL A. C., BAILEY T. C., DIGGLE P. J. and ROWLINGSON B. S. (1996) Spatial point pattern analysis and its application in geographical epidemiology. *Transactions, Institute of British Geographers*, **21**, 256–274.
GATRELL A. C. and SENIOR M. L. (1998) GIS and health. In Longley P., Maguire D., Goodchild M. F. and Rhind D. W. (eds) *Geographical Information Systems: Principles and Applications*, Geoinformation International, Cambridge.
HAYNES R. M., LOVETT A. A., GALE S. H., BRAINARD J. S. and BENTHAM C. G. (1995) Evaluation of methods for calculating census health indicators for GP practices. *Public Health*, **109**, 369–374.

Jones K. and Duncan C. (1995) Individuals and their ecologies: analysing the geography of chronic illness within a multilevel modelling framework. *Journal of Health and Place*, **1**(1), 27–40.

Kelsall J. E. and Diggle P. J. (1995) Nonparametric estimation of spatial variation in relative risk. *Statistics in Medicine*, **14**, 2335–2342.

Kingham S. (1993) Air pollution and respiratory disease in Preston: a geographical information systems approach, unpublished PhD, Department of Geography, Lancaster University.

Langford I. (1994) Using empirical Bayes estimates in the geographical analysis of disease risk. *Area*, **26**, 142–149.

Morris R. and Carstairs V. (1991) Which deprivation? A comparison of selected deprivation indexes. *Journal of Public Health Medicine*, **13**, 318–326.

Oliver M. A. et al. (1992) A geostatistical approach to the analysis of pattern in rare disease. *Journal of Public Health Medicine*, **14**, 280–289.

Openshaw S., Charlton M., Wymer C. and Craft A. W. (1987) A Mark 1 geographical analysis machine for the automated analysis of point data sets. *International Journal of Geographical Information Systems*, **1**, 335–358.

Riise T. et al. (1991) Clustering of residence of multiple sclerosis patients at age 13 to 20 years in Hordaland, Norway. *American Journal of Epidemiology*, **133**, 932–939.

Rowlingson B. S. and Diggle P. J. (1993) SPLANCS: spatial point pattern analysis code in S-Plus. *Computers and Geosciences*, **19**, 627–655.

Sabel C. E. and Gatrell A. C. (1998) Exploratory spatial data analysis of motor neurone disease in North West England: beyond the address at diagnosis. In Gierl L. et al. (eds.) *Proceedings of the International Workshop on Geomedical Systems*. Teulener-Verlag Stuttgart, Leipzig.

Scrivener G. and Lloyd D. C. E. F. (1995) Allocating census data to general practice populations: implications for study of prescribing variation at practice level. *British Medical Journal*, **311**, 163–165.

Selvin S., Merrill D. W. and Sacks S. (1988) Transformations of maps to investigate clusters of disease. *Social Science and Medicine*, **26**, 215–221.

Sooman A. and Macintyre S. (1995) Health and perceptions of the local environment in socially contrasting neighbourhoods in Glasgow. *Health and Place*, **1**, 15–26.

Tierney L. (1990) *LISP-STAT: An Object-Oriented Environment for Statistical Computing and Dynamic Graphics*. John Wiley, Chichester.

CHAPTER TWO

GIS as an Enabling Technology

GEOFFREY M. JACQUEZ

2.1 Introduction

An enabling technology is an instrument or technique that leads to an increased understanding of the natural world. Analogies have been drawn between GIS and enabling technologies that preceded fundamental scientific breakthroughs such as the optical microscope, the telescope and the electron microscope. In the 1600s Leeuwenhoek's water drop lens afforded the first images of bacteria (Dobel, 1960), leading, eventually, to a revolution in our understanding and treatment of infectious diseases. For environmental epidemiologists the promise of GIS is an increased understanding of links and perhaps causative relationships between environmental exposures and human health. But is this vision reasonable and attainable? This chapter explores this question by placing GIS within the context of a systematic approach to hypothesis generation and testing.

To set the context for the discussion we begin with a brief review of how GIS is currently used to analyse public health data. Next, the deductive approach to hypothesis testing of Popper is presented, along with the elaboration of Platt (1964). Under this approach, predictions are evaluated by experiment, an underlying theory is rejected when a prediction is falsified, and knowledge is increased in a systematic fashion, with new findings building on their predecessors. In practice science does not always advance in this way, yet the paradigm is useful because it encourages researchers to test their theories in an objective fashion.

Next we argue that hypotheses generated using GIS often are not expressed as falsifiable predictions, and, consequently, that the utility of these hypotheses is limited. While many published GIS studies are carefully crafted, others seem to suffer from a 'gee whiz' effect which precludes systematic hypothesis evaluation. This 'gee whiz' mode of investigation uses GIS to formulate visually striking thematic maps. After scrutinising the maps we derive hypotheses to explain apparent patterns, and, in the most egregious instances, educe health policy based at least partially on interpretation of these maps. This 'gee whiz' mode of investigation is invalid for several reasons: first, visual impact is a criterion used to select thematic maps for display, and the maps therefore are prone to displaying pattern where none exists; second, the maps are not subjected to statistical tests to determine whether apparent pattern is statistically significant and thus merits explanation; third, the derived hypotheses are of little value because they usually are not falsifiable. This chapter concludes by placing GIS and spatial statistics in the context of a more systematic approach that evaluates hypotheses emergent from GIS as falsifiable predictions.

2.2 Recent applications of GIS in health

While GIS as a technology is well developed, we are only now beginning to understand its implications for studies of human health. Considerable contributions are being made by GIS in exposure assessment, the identification of study populations, disease mapping, public health surveillance and the targeting of health interventions. This brief survey of GIS applications is not exhaustive, and the reader may wish to refer to other reviews of GIS in epidemiology (such as that by Clarke et al., 1996).

2.2.1 Exposure assessment

One area of great promise is exposure assessment, which ultimately seeks to reconstruct an individual's exposure to factors related to human health. In the United States the Agency for Toxic Substances and Disease Registry uses GIS-based exposure assessment to assess public health threats near hazardous waste sites (Agency for Toxic Substances and Disease Registry, 1992). Specific techniques include measured environmental data analysis, and environmental transport modelling (Holm et al., 1995), both of which require a GIS component. Measured environmental data analysis uses map interpolation techniques to estimate exposure. Environmental transport modelling estimates exposure using diffusion, wind-borne transport, hydrologic flow and other models of transport processes. While legitimate concerns have been raised regarding our ability to assess exposures at the individual level, particularly for diseases with long latencies (Marbury, 1996), the ability of GIS to link environmental data to residential histories is a promising direction. This was illustrated in the Hjalmars et al. (1994) study, which accounted for migration in their statistical analysis of childhood leukemia in Sweden in relation to the Chernobyl incident. Retrospective assessment of lead exposure in children based on residential history and housing data is another excellent example because the presence or absence of lead-based paints is strongly correlated with the age of the home (Wartenberg, 1992).

2.2.2 Identification of study populations

Several researchers have used GIS to identify subjects in case-control and cohort studies. Together, GIS's spatially referenced relational databases, address matching capabilities, buffering and topological operations allow potential exposures, covariates and confounders to be accounted for when identifying study subjects. Once identified, the study population is often analysed using non-spatial epidemiologic methods. Recent examples include a study of EMF exposure near power transmission lines in relation to leukemia (Wartenberg et al., 1993), in which GIS made the identification of study subjects possible, and the characterisation of residential demography in proximity to toxic waste sites (Croner and Sperling, 1996). It should be noted that GIS is useful for identifying study populations even when the disease outcome of interest does not have an inherently spatial component. In this regard the 'value added' of GIS arises from its ability to bring together diverse data through spatial referencing.

2.2.3 Disease mapping

While a GIS is not required to construct disease maps, it greatly eases this cartographic task. Because of its embedded relational database, GIS can more readily account for covariates during map construction. For example, Waller and McMaster (1997) propose a method for routine standardisation of disease rates, and illustrate their approach using GIS generated maps of leukemia in New York state. Bayes and empirical Bayes techniques for modelling disease rates allow one to account for spatial autocorrelation, covariates, and sources of error during map construction (Devine *et al.*, 1994; Bernardinelli and Pascutto, 1996; Xia and Carlin, 1997), but as yet these techniques are not commonly available in GIS.

2.2.4 Public health surveillance

Public health surveillance activities include the detection of epidemics, the assessment of infectious disease potential, and the design and evaluation of health interventions, among other activities (Teutsch and Churchill, 1994). Some authors foresee the evolution of a new field of public health informatics, which recognises a role for GIS in the management and analysis of public health surveillance data. Examples include the monitoring of reproductive outcomes in mothers living near hazardous waste sites (Stallones *et al.*, 1992). In addition, GIS is proving an invaluable tool in locating health care facilities and for targeting health care interventions (Barnes and Peck, 1994). Because of its ability to identify and map environmental factors associated with disease vectors, GIS is increasingly important in infectious and vector-born disease surveillance. Examples include Lyme disease (Glass *et al.*, 1995), malaria (Kitron *et al.*, 1994), and onchocerciasis (Richards, 1993), among others.

2.3 A philosophical perspective on the quest for knowledge

Notwithstanding these applications and others like them, some authors raise legitimate questions regarding the value of GIS in health. Marbury (1996, p. 89), notes that, for the most part, 'advances in environmental epidemiology will require carefully designed studies of rigorously defined outcomes combined with good measurements of personal exposure. It would be a shame to be distracted from this effort by the availability of a new tool that affords no new insights'. This reflects an expectation for GIS beyond its utility in exposure assessment, disease mapping and public health surveillance. Indeed, the optical microscope analogy raised earlier suggests that GIS, as an enabling technology, should lead to fundamental advances in our understanding of health–environment relationships. A key question then emerges:

Can we use GIS to *formulate* and *test* hypotheses in environmental epidemiology? And, if this is possible, how might it advance environmental epidemiology as a body of knowledge?

The philosopher Karl Popper (O'Hear, 1996) described a systematic approach to the accumulation of scientific knowledge (Figure 2.1) that sheds some light on this question. A theory or hypothesis is created to explain relationships or patterns in observed data. Testable predictions are then deduced from the theory and subjected to experiments to falsify or reject the prediction. If the prediction is rejected the theory is declared incorrect; if it is not rejected additional predictions are formulated and evaluated to further corroborate the theory.

Important lessons of this approach are first, that predictions can be falsified by experiment, but they cannot be proven. Second, theories can be rejected, but not proven. Third, to be useful, predictions must be falsifiable by experiment or some other mechanism. Predictions that cannot be falsified are useless. Finally, data from experiments designed to evaluate predictions must be independent of the original observations used to formulate theory. Testing predictions on data that gave rise to them is not valid.

John Platt's (1964) 'Method of Strong Inference' is an elaboration on the Popperian paradigm. Platt recognised that, to be useful, the Popperian approach to falsification of predictions must be undertaken in a systematic fashion within the context of a universe of plausible alternative hypotheses. Sir Arthur Conan Doyle's fictitious detective Sherlock Holmes solved apparently insolvable crimes using a similar approach: carefully enumerate the possible explanations, reject those that are falsified, and the remaining explanation, no matter how implausible, is the solution. Strong inference is illustrated in Figure 2.2.

First, the complete set of alternative hypotheses is enumerated. Experiments are then designed systematically to test the hypotheses. The experiments are then conducted, and the false hypotheses are winnowed out. The remaining hypothesis or

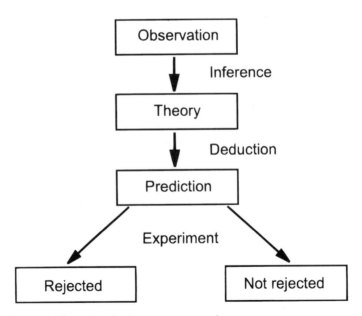

Figure 2.1 Schematic illustrating the Popperian approach.

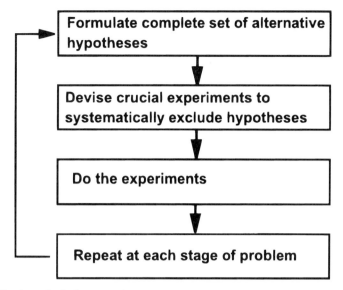

Figure 2.2 Platt's method of strong inference.

group of hypotheses are the viable explanations. Obviously, the set of alternative hypotheses may change as knowledge is gained during the experimental process. Platt's method of strong inference is a systematic approach for reaching the correct inference.

Some (for example, Koopman, 1996) criticise this as a simplistic process for making yes or no decisions about individual causal hypotheses. They point out that the assumptions underlying all theories and models are, at one level or another, incorrect, and that causal inference may therefore be precluded. It is undoubtedly true that science, as a system of knowledge, moves forward in many ways that include leaps of intuition and fortuitous circumstance, as well as by systematic hypothesis evaluation. Feyerabend (1981) enumerates several ways of evaluating scientific knowledge, and identified criticism, proliferation and realism as key ideas that play an ongoing and important role in the history of science.

Criticism means we do not simply accept phenomena (be they observations or larger systems of knowledge) as they are, but instead examine each one carefully and with jaundiced eye. Proliferation describes our tendency to work with a plurality of ideas, rather than single notions. This means, for instance, that we might recognise the contributions of a systematic approach, as well as those of fortuitous circumstance (which have a long tradition in science, including the discovery of penicillin via the wind-borne contamination of a bacterial culture by bread mold spores), with equal validity. Finally, realism recognises that our frameworks for dealing with scientific knowledge are themselves determined by the real world, and are not simply schemes for the mechanical processing of events.

From this perspective, arguments as to whether or not one approach or another is more or less the dominant paradigm are not particularly relevant, because our use of such paradigms is inherently eclectic. The deductive approach of Popper, and its elaboration as Platt's method of strong inference, is but one way of gaining scientific knowledge, and, in the context of this chapter, leads to useful conclusions regarding

GIS, health, and the accretion of scientific knowledge. An important contribution of strong inference is that, because it requires falsifiable predictions, it discourages the use of vague theories that are inherently not testable.

2.4 The 'gee whiz' effect

Environmental epidemiology seeks to elucidate the causes and correlates of disease in the context of a population's exposures as encountered in the everyday environment. When exploring spatial data with a GIS we often spend a lot of effort looking for spatial association among variables. In practice a deductive approach is not strictly followed, and each of us can probably think of studies that search only for association or corroboration, and do not evaluate falsifiable predictions.

GIS provides a powerful tool for managing spatially referenced data describing human health outcomes (for example, place of residence of patients with birth defects, cancers and so on), putative exposures (for example, groundwater contaminant concentrations, air pollution plumes, etc.) and their spatial relationships. After assembling the GIS database, one of the first steps in the hunt for associations between health and the environment is the creation of thematic maps (Figure 2.3). Thematic mapping involves the use of surface operations across coverages to generate summary maps. These operations, for example, might be used to identify areas where air and water quality are poor, and where birth defects are high. Thematic maps thus are useful for visualising spatial relationships among several variables. In fact, the perception of striking pattern is often the principal criterion that determines which map we retain for display to our colleagues and for continued analysis. The retained map is then inspected, and striking patterns on the map often prompt us to formulate explanatory hypotheses. This feat of unsupported inference is the 'gee whiz' effect, which I define as the formulation of hypotheses to explain an apparent pattern whose existence has not been confirmed (Figure 2.4).

Most of the studies described in the examples cited earlier use spatial statistics or related pattern recognition techniques to determine whether the perceived map

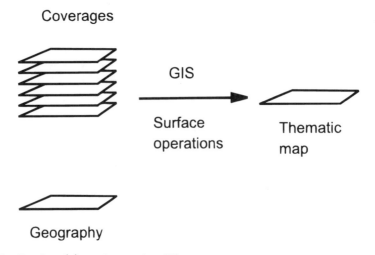

Figure 2.3 Creation of thematic maps in a GIS.

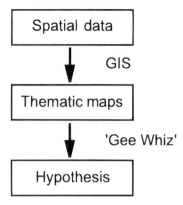

Figure 2.4 The 'gee whiz' mode of investigation.

pattern is real. Other examples of this genre include an analysis of breast cancer in the northeastern United States, which used a new spatial scan statistic to determine whether breast cancer cases cluster (Kulldorff and Feuer, 1997), and a study of birth defect rates in Des Moines, Iowa, which employed spatial Monte Carlo methods for assessing how unusual the map pattern really is (Rushton and Lolonis, 1996). GIS is increasingly used by public health professionals who are not familiar with such sophisticated statistical techniques and who do not have access to the requisite specialised software for the spatial-statistical analysis of mapped data. In these instances the GIS maps may not be analysed to determine whether their patterns are statistically unusual, and are prone to suffer, to a greater or lesser extent, from the 'gee whiz' effect.

This in large part is attributable to two causes. First, most GIS lack spatial statistical capabilities, and, conversely, most standard statistical packages lack methods for dealing with spatial data. Thus tools for statistically analysing GIS maps are not readily available. Second, the 'value added' of spatial-statistical analyses of GIS data is not widely known among health professionals, and hence the motivation for using spatial statistics is lacking. Here I argue that a major advantage of spatial statistical analysis of GIS data is that it helps avoid the unsupported inference of the 'gee whiz' effect (but also see the chapter by Gerard Rushton in this volume for additional applications illustrating the value added by GIS).

The 'gee whiz' mode of investigation is invalid for several reasons: first, visual impact is a criterion used to select thematic maps for display, and the maps therefore are prone to displaying pattern where none exists. Second, the maps are not subjected to statistical tests to determine whether apparent pattern is statistically significant and thus merits explanation. Third, the derived hypotheses are of little value because they usually are not falsifiable. Before we can formulate a hypothesis to explain perceived patterns on a map, we first must determine whether the perceived pattern is in some sense unusual, and therefore warrants an explanation.

2.5 GIS and spatial statistics

This returns us to the question raised earlier. Can we use GIS to *formulate* and *test* hypotheses in environmental epidemiology? The Popperian approach provides

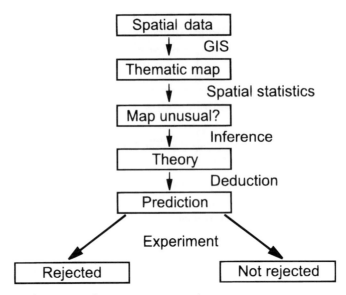

Figure 2.5 GIS in the context of a systematic approach.

some insight to this question (Figure 2.5). I consider the three steps in a GIS; spatial data, map creation, and determining whether the map is unusual, to be equivalent to the observations in this Popperian paradigm. That is, we use the GIS and spatial statistics as a 'lens' through which we observe the world. GIS surface operations are used to generate thematic maps, and spatial statistics determine whether pattern on these maps is in some sense unusual. Useful spatial statistics for accomplishing this include disease clustering methods, methods for analysing spatial point distributions, adjacency statistics for determining whether classes of areas share common borders, tests for boundary overlap, and statistics for evaluating association between two or more spatial variables. In this volume Haining describes spatial statistical methods suited for detecting associations between health and socioeconomic data at the census unit level, and Kulldorff (1996) reviews several disease clustering tests for analysing spatial point patterns describing the locations and times of occurrence of health events (but also see Jacquez et al., 1996a, b, for a recent review of disease cluster tests). The ability of GIS to manage detailed spatially referenced data from several variables simultaneously supports statistical methods that operate on samples from a data surface. Here a data surface is the continuous geographic distribution of a variable's value, and is appropriate when a variable could be observed anywhere in the study area. An example is the data surface formed by an air pollution plume. Jacquez (1995) describes statistical techniques for the analysis of geographic boundaries, which are defined by zones of rapid change in a data surface. The techniques mentioned above are useful for determining whether a map's spatial pattern is statistically unusual.

Once a map is deemed to be statistically unusual, we are justified in inferring a theory or hypothesis to explain the spatial relationships. We then formulate a testable, meaning a falsifiable, prediction, and design an experiment to test that prediction. Again, this approach has power only to falsify, not prove, the theory.

At least three kinds of experiments seem possible. We may design an *epidemio-*

logic study to test predictions describing disease occurrence in populations. A *laboratory study* may be designed when the prediction describes disease progression at the organismic level. Finally, another *GIS study* may be used to evaluate epidemiologic predictions that involve a spatial dimension.

In practice I suspect GIS studies will not be very useful for evaluating spatial predictions, for two reasons. First, spatial systems are difficult to manipulate, and for this reason designed experiments will often be impossible or very difficult. Experimentation on human populations is not acceptable, and for this reason one cannot vary pollution output to determine that point at which exposures have acute or chronic effects. In some instances natural experiments may be available through inadvertent circumstance, and should be exploited. For example, the exposure of populations to caesium-137 and other radionucleides from the Chernobyl accident is an unintended experiment that afforded an assessment of the low-dose effects of ionising radiation (Hjalmars *et al.*, 1996).

Second, GIS data used to formulate theory should not be used to test predictions emergent from that theory. To do so would bias one towards confirming the observed pattern. This means, of course, that one cannot evaluate hypotheses emergent from a GIS study using the same GIS database. Kulldorff and Feuer (1997) refer to this as 'pre-selection bias' in their study of breast cancer clusters in the northeastern United States.

An additional problem is the spatial uncertainty inherent in most epidemiologic data. This uncertainty has several sources. It emerges when centres of areas such as zip-code zones or census tracts are used instead of exact place of residence. Uncertainty also arises when data are gridded (as, for example, in raster-based GIS) and the coordinates of the nearest grid node are used instead of exact locations. Conceptually, locations are almost always uncertain because humans are mobile rather than sessile and because health events and their causative exposures may occur anywhere within a person's activity space. Tobler *et al.* (1995) observed that in modern society a person's daily activity space is approximately 15 km and varies widely. Exact locations do not represent the mobility of modern societies.

There are many examples of uncertain locations in the literature. Cuzick and Edwards (1990) used the centres of postal code zones to represent place of residence in their study of childhood leukemia in North Humberside. Waller *et al.* (1995) mapped locations of cases of childhood leukemia in Sweden at parish centroids, and evaluated proximity to nuclear power plants using several methods. Location uncertainty arises in these two examples because area centroids are used instead of exact locations. In a study notable for its spatial resolution, Lawson and Williams (1994) used place of residence to assess a possible cluster of deaths from respiratory cancer near a smelter in Armadale, Scotland. Indoor air quality is known to differ substantially from that outdoors, and an unknown proportion of each individual's exposure to smelter fumes presumably occurred outside of the home. In this example location uncertainty arises because place of residence is used to represent exposures that occurred throughout each person's activity space. Kulldorff and Fever (1997) used the latitude and longitude of county centroids in their assessment of breast cancer clusters, and this practice is encouraged in a recent paper by Croner and Sperling (1996) who suggest that centroids are frequently a revealing way to manipulate data sets. This approach, however, uses the centroid as the single coordinate to which averages, counts and other data attributes for an area are assigned, and the results will depend, more or less, on one's choice of each centroid's coordinate value.

These studies illustrate the ubiquity of the location uncertainty problem. In practice geographic locations are particularly uncertain when they are used as proxy measures of exposure. If we had more information we could explore dose–response relationships using controlled studies. When such detailed knowledge is lacking it is not uncommon to use spatial relationships among cases, and their proximity to putative sources of hazard, as an uncertain measure of exposure. Many spatial statistical techniques used for analysing mapped health data assume exact locations. For example, point-based disease cluster statistics, such as Mantel's (1967) and Cuzick and Edwards' (1990) test, assume location data of the form (x, y), where x, y is a precise spatial coordinate of a case. Applying such techniques to centroid locations can produce misleading results (Jacquez and Waller, 1996). Appropriate approaches for dealing with location uncertainty need to be developed to improve the resolution of our spatial statistical analyses of health data from GIS.

Returning to the original question, can GIS be used to formulate and test hypotheses in environmental epidemiology? I think the primary contribution of GIS will be in the formulation of hypotheses to explain statistically significant spatial relationships between the environment and human health. However, because of difficulties in manipulating spatial systems, GIS will be less useful in testing predictions emergent from these hypotheses.

2.6 Conclusion

This chapter was motivated by the observation that our current use of GIS, much like the early use of the water lens, is driven primarily by the technology itself, rather than by a scientific method. At first the water drop lens was a novelty with the 'gee whiz' cachet of providing a window on the world of Leeuwenhoek's 'animalcules'. The compound microscope using glass lenses was devised in the 1600s, and ultimately enabled Pasteur and others in the 1800s to recognise the association between bacteria and infection. But it was only after Koch placed this technology in the context of a systematic approach for elucidating disease causation that its promise for increasing our understanding of disease processes was fully realised. (I have taken liberty in summarising the history of microscopy, readers may wish to consult Locy (1925) for a more accurate account.)

We are at an early stage in the development and use of GIS in environmental epidemiology, analogous, perhaps, to the first applications of the compound microscope; the technology is well developed, but where it will lead in terms of increasing our understanding of relationships between health and environment is still unclear.

So what is the lesson of this chapter? Appreciate that we are at an early stage in our use of GIS, and that the excitement of the 'gee whiz' inferential leap plays an important role in stimulating us to explore this technology's capabilities. Be cautioned, however, against the 'gee whiz' mode of investigation, which encourages unbridled speculation and gives free reign to the ability of the human eye to see pattern where none exists, for this may lead to false findings and will ultimately undermine the field's credibility. Finally, recognise that the full promise of this technology will not be realised until we couch hypotheses emergent from GIS as falsifiable predictions within a systematic framework. In my opinion this field is still waiting for its Pasteurs and Kochs to apply GIS within the context of a systematic approach.

Acknowledgments

This research was funded in part by Small Business Innovation Research grant R43 CA65366-01A1, and Small Business Technology Transfer grant R41 CA64979-01 from the National Cancer Institute. Its contents do not necessarily represent the official views of the NCI.

References

AGENCY FOR TOXIC SUBSTANCES AND DISEASE REGISTRY (1992) *Public Health Assessment Guidance Manual*. Lewis Publishers, Chelsea and Michigan.

BARNES S. and PECK A. (1994) Mapping the future of health care: GIS applications in health care analysis. *Geographic Information Systems*, **4**, 31–33.

BERNARDINELLI L. and PASCUTTO C. (1997) Disease mapping with errors in covariates. *Statistics in Medicine*, **16**, 741–752.

CLARKE K. C., MCLAFFERTY S. L. et al. (1996) On epidemiology and geographic information systems: A review and discussion of future directions. *Emerging Infectious Diseases*, **2**, 85–92.

CRONER C. and SPERLING J. (1996) Geographic Information Systems (GIS): New perspectives in understanding human health and environmental relationships. *Statistics in Medicine*, **15**, 1961–1977.

CUZICK J. and EDWARDS R. (1990) Spatial clustering for inhomogeneous populations. *Journal of the Royal Statistical Society*, Series B, **52**, 73–104.

DEVINE O. J., LOUIS T. A. et al. (1994) Empirical Bayes estimators for spatially correlated incidence rates. *Environmetrics*, **5**, 381–398.

DOBEL C. (1960) *Antony van Leeuwenhoek and his 'Little Animals'*. Dover Publications, London.

FEYERABEND P. K. (1981) *Realism, Rationalism and Scientific Method*. Cambridge University Press, Cambridge.

GLASS G. E., SCHWARTZ B. S. et al. (1995) Environmental risk factors for Lyme disease identified with geographic information systems. *American Journal of Public Health*, **85**, 944–948.

HJALMARS U., KULLDORFF M. et al. (1994) Risk of acute childhood leukemia in Sweden after the Chernobyl reactor accident. *British Medical Journal*, **309**, 154–157.

HJALMARS U., KULLDORFF M. and GUSTAFFSON G. (1996) Childhood leukemia in Sweden: Using GIS and a spatial scan statistic for cluster detection. *Statistics in Medicine*, **15**, 707–715.

HOLM D. M., MASLIA M. L. et al. (1995) Geographic information systems: A critical resource in exposure assessment. *SUPERFUND XVI Conference and Exhibition*, Washington, D.C.

JACQUEZ, G. M. (1995) The map comparison problem: Tests for the overlap of geographic boundaries. *Statistics in Medicine*, **14**, 2343–2361.

JACQUEZ G. M., GRIMSON R., WALLER L. and WARTENBERG D. (1996a) The analysis of disease clusters Part II: Introduction to techniques. *Infection Control and Hospital Epidemiology*, **17**, 385–397.

JACQUEZ G. M. and WALLER L. A. (1996) The effect of uncertain locations on disease cluster statistics. *Second International Symposium on Spatial Accuracy Assessment*, Fort Collins, Colorado.

JACQUEZ G. M., WALLER L., GRIMSON R. and WARTENBERG D. (1996b) The analysis of disease clusters Part I: State of the art. *Infection Control and Hospital Epidemiology*, **17**, 319–327.

KITRON U., PENER H. et al. (1994) Geographic information system in malaria surveillance: mosquito breeding and imported cases in Israel, 1992. *American Journal of Tropical Medicine and Hygiene*, **50**, 550–556.

KOOPMAN J. (1996) Epidemiology seen more broadly. *Epidemiology Monitor*, Roswell, Georgia, pp. 5–6.

KULLDORFF M. (1996) Statistical Methods for Spatial Epidemiology: Tests for Randomness. In Gatrell A. and Löytönen M. (eds) *GIS and Health*, Taylor and Francis, London, 49–62.

KULLDORFF M. and FEUER E. J. (1997) Breast cancer clusters in northeastern United States: A geographical analysis. *American Journal of Epidemiology*, **146**, 161–170.

LAWSON A. B. and WILLIAMS F. L. R. (1994) Armadale: A case-study in environmental epidemiology. *Journal of the Royal Statistical Society*, **157**, 285–298.

LOCY W. A. (1925) *The Story of Biology*. Garden City Publishing Company, New York.

MANTEL N. (1967) The detection of disease clustering and a generalized regression approach. *Cancer Research*, **27**, 201–220.

MARBURY M. (1996) GIS: New tool or new toy? *Health and Environment Digest*, **9**, 88–89.

O'HEAR A. (1996) *Karl Popper, Philosophy and Problems*. Cambridge University Press, Cambridge.

PLATT J. R. (1964) Strong inference. *Science*, **146**, 347–353.

RICHARDS F. O. (1993) Uses of geographic information systems in control programs for onchocerciasis in Guatemala. *Bulletin of the Pan American Health Organization*, **27**, 52–55.

RUSHTON G. (1996) Instructional modules on a CD-ROM for improving public health using GIS. In Gatrell A. and Löytönen M. (eds) *GIS and Health*, Taylor and Francis, London, 63–79.

RUSHTON G. and LOLONIS P. (1996) Exploratory spatial analysis of birth defect rates in an urban population. *Statistics in Medicine*, **15**, 717–726.

STALLONES L., NUCKOLS J. R. et al. (1992) Surveillance around hazardous waste sites: GIS and reproductive outcomes. *Environmental Research*, **59**, 81–92.

TEUTSCH S. M. and CHURCHILL R. E. (1994) *Principles and Practice of Public Health Surveillance*. Oxford University Press, Oxford.

TOBLER W., DEICHMANN U. et al. (1995) *The global demography project*. National Center for Geographic Information and Analysis, Santa Barbara.

WALLER L. A. and MCMASTER R. B. (1997) Incorporating indirect standardization in tests for disease clustering in a GIS environment. *Geographical Systems*, to appear.

WALLER L. A., TURNBULL B. W. et al. (1995) Detection and assessment of clusters of disease: An application to nuclear power plant facilities and childhood leukemia in Sweden. *Statistics in Medicine*, **14**, 3–16.

WARTENBERG D. (1992) Screening for lead exposure using a geographic information system. *Environmental Research*, **59**, 310–317.

WARTENBERG D., GREENBERG M. et al. (1993) Identification and characterization of populations living near high-voltage transmission lines: A pilot study. *Environmental Health Perspectives*, **101**, 626–632.

XIA H., and CARLIN B. P. (1997) Hierarchical models for mapping Ohio lung cancer rates. *Environmetrics*, **8**, 107–120.

CHAPTER THREE

Spatial Statistics and the Analysis of Health Data

ROBERT HAINING

3.1 Introduction

Spatial epidemiology is the analysis of spatial and space–time distributions of disease data. Such analysis enables the identification of populations with high relative risks for particular diseases and may help to isolate possible causal factors for subsequent analysis by individual level study designs. Area studies may also be important in their own right in the case of certain diseases, such as respiratory, water-borne or those linked to exposure to radiation, where individual level epidemiological studies would be incapable of establishing accurate, individual exposure levels to critical risk factors. Area-based studies, particularly where similar results are found at different times and in different places may give aetiological clues.

Health services research is concerned with health promotion and disease prevention and focuses on questions to do with the need for, provision and use of, particular services as well as the effects of changes in service provision. As with spatial epidemiology, health services research may be directed at specific geographically defined populations seeking to identify and address questions arising from health inequalities in a population. For example, it addresses questions about how service provision impacts on different groups living in different parts of a service catchment thereby contributing to an understanding of the extent to which the health needs of specific populations are being met.

There is growing awareness of the role GIS can play in handling the large volumes of spatially referenced data routinely collected at small spatial scales in the field of public and environmental health (de Lepper *et al.*, 1994). At one level this can mean providing a facility that will help with the mapping and display of such data – although if goals remain as limited as this a mapping package is probably adequate. At a higher level the availability of GIS opens up the possibility for more detailed handling and interrogation of spatially referenced data including the types of questions cited above. However, as has been frequently mentioned (for example, Haining, 1994b) the analysis capability of GIS is quite limited and this in turn limits the extent to which GIS can become a general purpose tool for spatial analysis – in any field of research. The results of an earlier project developing GIS in the area of health needs assessment, and from which the work described in Section 3.5 grew, is reported in Haining (1996).

In this chapter we shall discuss some of the areas of spatial statistics that are important for spatial epidemiology and health services research. Following a description of some important spatial statistical models and techniques in Section 3.2 there will be a brief discussion in Section 3.3 of the problem of constructing an appropriate areal framework for analysis. Section 3.4 will discuss data quality and Section 3.5 will report on a project that is extending GIS capability to include spatial statistical analysis. Spatial data analysis requires the user to approach data in often quite novel ways, something that is not made any easier by a lack of appropriate software for implementing specialist techniques.

3.2 Spatial statistics

This section is divided into two main parts: methods for the exploratory analysis of spatial data and methods for analysing relationships including fitting models and testing hypotheses. Only methods for describing pattern and analysing relationships in area data, what Cressie (1991) calls 'lattice data', will be discussed. Formally this means that we assume that the study region (R) has been divided into a set of zones or areas (D) – for example a British city divided into enumeration districts or wards or a nation divided into subregions such as counties or health authority regions. Attached to each area ($i \in D$) is a random vector $\mathbf{Z}_i$ that describes the set of attributes attached to the area. The attributes may include not only medical attributes of the population but demographic, socioeconomic and environmental attributes.

3.2.1 Conceptual models of spatial variation: pattern detection and exploratory analysis

Exploratory spatial data analysis (ESDA) is a collection of statistically robust techniques for identifying different forms of spatial variation in spatial data. It represents the extension of exploratory data analysis (EDA) into the domain of spatial data (Hoaglin et al., 1983). An underlying data model for EDA on a data set for a single variable Z assumes two main components to the data:

$$\text{Data } (Z) = \text{smooth} + \text{rough} \tag{3.1}$$

and EDA techniques, which utilise visual and graphical tools as well as numerical measures, are designed to assist in the identification of these components. The 'smooth', sometimes called 'fit', comprises large scale regular features of the data on Z. The 'rough', sometimes called 'residuals', comprises small scale features of the data on Z that relate to individual data cases or small subsets of cases within the full data set. In the case of spatial data, where the analyst is seeking to describe spatial characteristics of a single variable in R, we shall define the 'smooth' component as regional scale or 'global' patterns in the data. One type of regional scale pattern is a trend the presence of which reflects the existence of an overall gradient in some disease such as, for example, the often commented on, south-east to north-west increase in heart disease mortality in England. In addition to this large scale gradient other 'smooth' properties may be present in the form of variation around the trend. Superimposed on the trend surface or gradient might be spatial covariation – adjacent areas tending to have similar levels of Z. Such a feature is also termed 'spatial autocorrelation'. As an illustration of these two 'smooth' components, asthma rates may decrease with increasing distance from the centre of a large conurbation, reflecting perhaps an overall decrease in air pollution away from the centre, and this might be represented as a trend surface. Superimposed on this general trend, however, might be spatial autocorrelation in asthma rates reflecting spatial covariation (spatial similarity) in air pollution levels or the general mobility of the urban population in their work and other travel patterns that bring them into regular contact with air pollution levels in areas adjacent to their residential area within the conurbation. In addition to these two global pattern properties representing 'smooth' properties of the surface there may be 'local' elements of the spatial pattern, for example individual or small clusters of areas with particularly high (hot spot) or low (cold spot) rates. These would represent 'rough' properties of the surface. Such 'local' elements of pattern in asthma rates might reflect pockets of gentrification near the city centre enjoying lower rates than the global elements of the surface because these households have the resources to enjoy a more healthy lifestyle, or pockets of deprivation in suburban council housing areas suffering higher rates than the global elements of the surface because these households have poorer housing conditions than the average suburban dweller. The underlying spatial data model thus comprises 'global' and 'local' scale patterns corresponding to the terms 'smooth' and 'rough' in the exploratory data model. In summary:

Data (Z) = (trend + spatial covariation)

    global or smooth

+ (individual or groups of hot (and cold) spot areas)

    local or rough

The partition into different forms of spatial variation described above is only one possible model for spatial variation in an ESDA framework. Getis and Ord (1992), for example, distinguish between global and local scales of variation and, in the context of a positive valued attribute with a natural origin, suggest a model that classifies a map on the basis of the degree of spatial concentration in either large or small attribute values. (In a later paper, Ord and Getis (1995), the requirement of a natural origin is removed.) In the terminology of (3.1) the smooth element is the existence of a general propensity for large (or small) values to be found together but within such a global picture there may exist smaller groups of areas of particularly high or low concentration and these would seem to equate with the rough element of (3.1). In a disease map, for example, there might be a general tendency for all areas with high death rates to concentrate together in one or more parts of the map (mainly 'smooth', little or no 'rough'), but at the other extreme there might be no evident concentration of similar sized rates except in a relatively small number of adjacent areas in one or two parts of the map where high rates are found together (little or no 'smooth', mainly 'rough'). The distinction between 'spatial covariation' in the first conceptual model and 'spatial concentration' in this does not appear to be sharp and it may be possible to reconcile these two conceptual (ESDA) models within a single more general model of spatial variation. However, whatever the model the description of spatial variation is dependent on the scale and nature of the spatial partition (D). The presence of spatial concentration or local clusters superimposed on larger scale gradients or trends will depend on the size of the areal units relative to the scale of the background variation responsible for the local scales of variation.

A number of techniques have been proposed to assist in identifying the 'smooth' and 'rough' elements of a spatial data set. A distinction is drawn between 'general' or 'global' tests, also sometimes called 'whole map' statistics, concerned with identifying overall regional patterns (the 'smooth') and 'focused' or 'local' tests which concentrate on individual areas, $i \in D$, (the 'rough'). This focus on the 'local' might be a sweep through all $i \in D$ to look for evidence of 'rough' wherever it might be found, or it might be concerned with only one or two areas (i) in D perhaps because they possess a special attribute which the analyst feels might have implications for health such as a nuclear plant or waste incinerator. See also the chapter by Kulldorff in this volume.

Cressie (1994) proposed median polish with row and column effects to identify gradients in a spatial dataset. His original application was to gridded data but Cressie and Read (1989) carried out a median polish on sudden infant death syndrome data for a set of counties in North Carolina. Unfortunately the method when applied to non-gridded data requires some *ad hoc* decisions to be taken to transform the irregular county data to a grid. Ord and Getis (1995) applied the G_i ($i \in D$) statistic to county level AIDS data in California to test for trends in the incidence of AIDS away from San Francisco. The G_i statistic measures spatial concentration and is an example of a focused or local statistic, it was computed here only for $i =$ San Francisco but taking successive distance bands at increasing distance from San Francisco. The plot of the resultant G_i values against distance declined with increasing distance from San Francisco providing evidence of a general decline in AIDS with distance from that city. Haining (1990) used a sequence of box plots computed over increasing distance bands from the centre of Glasgow to show the presence of trend in standardised mortality rates. The presence of spatial covariation can be explored in a variety of ways – rigorously by using an appropriate spatial autocorrelation statistic (Cliff and Ord, 1981), more informally, for example using a scatter plot and plotting each value against the mean of the neighbouring areas (Haining, 1990). In both cases the analyst may wish to first remove any spatial gradient in the data set. Ord and Getis (1995) construct the global G statistic which in the context of their model of spatial variation measures the general propensity for attribute values of similar size to concentrate together.

Testing for the presence of 'rough' elements of the data might be undertaken by looking for spatial outliers or even clusters of outliers after removing the trend and spatial covariation elements in the data. Suppose trend has been extracted from a spatial data set and the resultant data set plotted by taking each value and plotting it against the mean of its neighbouring areas. A simple way of signalling outliers is to run a least squares regression through the scatter. Outliers from the regression are indicative of spatial outliers from these 'smooth' properties of the data (Haining, 1990, pp. 197–227). Anselin (1995) has recently drawn attention to LISA's (Local Indicators of Spatial Association) which are statistics for picking up local spatial properties. These properties can include local concentrations of events detected by the G_i and G_i^* statistics of Getis and Ord (1992), or local patterns of covariation detected by variogram clouds, pocket plots (Cressie, 1994; 1991) and local Moran plots (Anselin, 1995). There is the potential to generate enormous numbers of diag-

Table 3.1 A summary of components of spatial variation and corresponding techniques

Global/'smooth'		Local/'rough'	
Model component	Technique	Model component	Technique
Trend	Median polish G_i test Box plots	Hot spots and cold spots	Residuals outlier tests
Spatial covariation	Moran test Geary test Scatter plots	Local spatial covariation	Cloud plots Pocket plots Local Moran plots
Concentration	G test	Local concentrations	G_i, G_i^* tests

nostic statistics for any set of N areas and it is perhaps appropriate to add a warning note about generating a large number of statistical estimates based on very small subsets of the data (see, for example, the warning in Cressie, 1994). The comparable statistical methodology seems to be that concerned with using the evidence to assess data influences on estimating statistical models or identifing outliers but this is based on deleting individual cases from the full data set rather than the other way round. Also if the user has no explicitly articulated model for the spatial variation it may not be clear what the measures signify about the spatial distribution of values. Table 3.1 provides a summary of techniques in relation to these conceptual models.

3.2.2 Mathematical models of spatial variation: model specification

Rigorous methods based on explicit models of the data are also available. Large scale gradients can be represented by trend surface models. These are linear or higher order functions in the spatial coordinates of the area. The location of each area might for example be represented by the X,Y coordinates of its (population or geometric) centroid. Such models can be augmented by the addition of terms that capture the spatial autocorrelation in the residuals from the trend surface. Spatial autocorrelation models typically represent the value of an attribute (Z) at location i (Z_i) as some function of the values of the same attribute at the neighbours of i ($N(i)$). There are many such models and a discussion of these is beyond the scope of this review but the interested reader can obtain an extended discussion in Haining (1990, pp. 65–117, 249–282) which also discusses the technical problems associated with their fitting. However by way of example a linear trend surface model with autocorrelated errors capturing the two components mentioned above for a single attribute Z might be represented as:

$$Z_i = \beta_0 + \beta_1 X_i + \beta_2 Y_i + u_i$$
$$u_i = \rho \Sigma_{j \in N(i)} u_j + e_i \qquad (3.2)$$

To revert briefly to the earlier terminology, in this model the terms in X and Y represent the large scale linear gradient (with parameters β_0, β_1 and β_2) and the term in u represents the spatial autocovariance around the trend (with parameter ρ).

Together these form the 'smooth'. The term in e is the residual, or 'rough', and is an independent and identically distributed noise process. There is a brief introduction to these models in Richardson (1992).

3.2.3 Analysing relationships in geographical populations

Testing for relationships between, for example, disease events or uptake levels of a service and population and environmental characteristics involves the use of correlation and regression. While at first sight these might seem familiar statistical tools, their application to geographical data poses a number of problems. These problems are both interpretative, stemming from the fact that relationships refer to aggregates of individuals, and technical, stemming from the properties of the underlying random process generating Z.

Problems of interpretation fall into several categories. The *ecological fallacy* (or *ecological bias*) is the difference between estimates of relationships at the aggregate level from those at the individual level. Awareness of this is important if we are not to overstate the strength of any relationship in the individuals of a population that has been observed in the spatially aggregated data. Further, the particular form of the spatial aggregation can also affect the estimate of the relationship and is referred to as the *modifiable areal units problem*, the term stemming from the fact that areal units are not 'natural' but usually arbitrary constructs. This latter problem contains two effects: one effect derives from holding the scale of the aggregation constant but grouping different individuals together – effectively selecting different areal boundaries while holding the overall size and number of areal units constant; the other effect derives from reducing the number but increasing the size of the areal units. The latter might more properly be called a 'scale effect'. It has been noted that as the size of areal units increases, and hence their number decreases, the measure of association tends to increase. The effect of increasing areal unit size is bound up with loss of variability or smoothing in the data induced by the aggregation process. In the case of bivariate correlation this variance appears in the denominator of the statistic which is why this measure of association increases. For example if the population attribute is the Townsend index of deprivation then as larger and larger areal units are used, then the area will tend to contain a population with a greater and greater mixture of deprivation levels. The deprivation score for the area becomes less and less representative of the population in the same way that the mean is a very limited descriptor of a frequency distribution if that frequency distribution possesses a large spread. However, if the analyst is tempted to draw the conclusion from this that therefore it is better to work with small regions to try to ensure homogeneity of the population variables this runs up against a counter problem. Rate estimation, such as computing the standardised incidence rate, is more reliable when computed for large populations than small. Where population counts are small, the addition or subtraction of a few cases will have a far greater effect on computed rates than in the case where population counts are much larger. This is true for any type of rate estimation but is particularly true in the case of relatively rare diseases where the occurrence of a small number of cases can have a large impact on computed rates. There is clearly something of a trade off required here – areas with large enough populations to generate reliable rates but homogeneous with respect to those factors the analyst wishes to explore as possibly helping to explain those rates.

There is a further problem in the case of testing for relationships in spatial data which is the problem of *spurious correlation*. A particular form of this problem arises when 'potential confounder variables ... show the same regular spatial pattern' (Richardson, 1992, p. 183). For example, in trying to disentangle the contribution of an environmental influence such as air pollution from deprivation associated with poor housing conditions on respiratory disease rates it tends to be the case that the most deprived groups often live in the centres of cities where air pollution may also be greatest. In such circumstances it may be useful to investigate any relationship both before and after the removal of a gradient particularly if it is suspected that the gradient may reflect the influence of confounding variables. An example is described in Haining (1991a).

Technical problems, particularly in analysing relationships between, for example, disease rates and deprivation levels stem from two properties of such data. As described in the previous section values for either or both disease and deprivation may be *spatially autocorrelated*. This may arise, for example, because the continuity of attribute characteristics is at a scale larger than the areal units. It may also arise because the underlying random process is responsible for generating similar levels in adjacent areas as in the case of an infectious disease for example. In the presence of spatial autocorrelation, while the construction of the Pearson correlation coefficient remains unchanged the inference theory is greatly altered (Clifford and Richardson, 1985; Clifford et al., 1989). Allowance has to be made for the presence of spatial autocorrelation in the two variables and this is done by computing the 'effective sample size' based on estimates of that spatial autocorrelation. Haining (1991b) discusses implementation of the method with a medical application and extends the findings to the Spearman rank correlation coefficient.

The presence of spatial autocorrelation in the residuals (strictly speaking the errors) of a regression model violates one of the statistical assumptions underlying least squares regression and may result in invalid inferences. The estimates of the sampling errors of the parameters of the regression model are underestimated when residuals are positively autocorrelated. As a result independent variables may be retained in the model as significant when they are not (type 1 error). The coefficient of multiple determination, measuring the goodness of fit of the model, is inflated. The underlying cause of residual autocorrelation is often the omission of independent variables from the model that are spatially autocorrelated and which have a significant influence on the variation in the disease rate under investigation. The Moran test is often used for testing for the presence of residual autocorrelation and if it is found to be present either the analyst must try to identify the missing independent variables or fit a regression with correlated errors model or some similar 'spatial regression' as described below. Unfortunately fitting procedures for these models are not routinely available in standard statistical software packages – nor indeed is the Moran test. There is an extended discussion of the regression model with autocorrelated errors in Haining (1990, pp. 123–129) and Haining (1994a).

Another commonly occurring technical problem is non-constant variance, heteroscedasticity, of regression residuals (again strictly speaking the errors). As before this undermines inference. The presence of heteroscedasticity arises from variation in the number of observed cases of the disease between regions. The underlying process generating the observed disease count in any area (i) is binomial. If the binomial parameter p_i, which is the probability that an individual catches the disease in area i goes to zero, and n_i, which is the number of individuals in area i, goes to infinity under the condition that their product converges to a finite constant (λ_i) then the observed count in area i is Poisson with parameter λ_i. Poisson regression can therefore be used to model the variation in observed counts which is then conceptualised as the outcome of sampling variation, explained variation associated with specified independent variables and error. Alternatively the counts can be converted to standardised rates, logarithm (to the base e) transformed and the resulting variable is then normally distributed with mean p_i and standard deviation which is a function of the reciprocal of the observed count in each area. The argument, described in detail in Pocock et al. (1981) and summarised in Haining (1991a), demonstrates that regression modelling may either be undertaken using Poisson regression (for which no test for residual spatial autocorrelation appears to exist nor a fitting procedure where spatial autocorrelation needs to be allowed for in the model) or by normal regression modelling but for which there is the additional problem of heteroscedastic error.

3.2.4 Special issues in analysing relationships in areal data

The previous section has illustrated that while standard techniques of correlation and regression can be called upon to analyse relationships in the case of area disease data there are problems of interpretation but also problems of a technical nature which make the use of these statistical methods difficult. However, there is a further issue which originates from an earlier comment about the fact that geographical areas are artificial constructs. Area boundaries, typically those used for analysis such as health authority regions, wards or enumeration districts, have no relevance to the process generating the disease counts. Put slightly differently the occurrence or non-occurrence of a disease is not influenced by the distribution of ward boundaries. For certain kinds of disease the open and permeable nature of areal unit boundaries is important. In the case of an infectious disease such as an influenza outbreak the rate of the disease in an area (i) may in part be a function of the rate of the disease in neighbouring areas ($N(i)$) stemming from the interpersonal communication between residents of nearby, even adjacent areas. This might lead to a model specification of the form:

$$Z_i = \beta_0 + \beta_1 X_{1i} + \beta_2 X_{2i} + \phi \Sigma_{j \in N(i)} Z_j + e_i \tag{3.3}$$

where Z is the standardised incidence rate of the infectious disease, X_1 and X_2 are independent variables, perhaps measuring socioeconomic characteristics, and the final term in the regression model is modelling the effect of rates in the set of neighbouring areas on the rate in i. The parameters of the model are $\beta_0, \beta_1, \beta_2$ and ϕ, and e_i is the independent error term. This model is called a spatial regression model with spatially lagged dependent variable. The model cannot be fit by ordinary least squares, unlike the equivalent time series version of this model, and the maximum likelihood procedure yields an estimator with better statistical properties (Haining, 1990).

In the case of certain kinds of disease induced by environmental factors a further variant of the usual regression model might be appropriate. Rates of respiratory disease in area i might be a function of levels of air pollution. Individuals resident in area i, as a result of their general patterns of mobility within the city, might be exposed to levels of the risk factor in areas other than just area i. Thus if Z is the standardised incidence rate of the environmentally induced condition (respiratory disease) and X_2 is the relevant environmental variable (air pollution) then this might lead to a model specification of the following form:

$$Z_i = \beta_0 + \beta_1 X_{1i} + \beta_2 X_{2i} + \gamma \Sigma_{j \in N(i)} X_{2i} + e_i \tag{3.4}$$

where X_1 is an independent variable where there are no spatial 'spillover' effects in the effect it has on Z, and the parameters of the model are β_0, β_1, β_2 and γ, and e_i is the independent error term. This is called a spatial regression model with spatially lagged independent variables. The model can be fit by least squares regression but does raise the problem of multicollinearity particularly if the variable that is lagged is itself spatially autocorrelated which in the case of many environmental variables is likely to be the case. It should be noted that in all the regression models ((3.2), (3.3) and (3.4)) both the spatial and non-spatial effects are 'whole map' effects in that the relationships are assumed to hold for all $i \in D$ and parameters are spatially invariant. Diagnostics exist to assess the extent to which individual areas do not appear to fit with this assumption, or affect parameter estimates in the case of (3.2) and (3.4) but not (3.3). (See, for example, Haining (1994a).)

The problem of measuring the association between two variables where data refer to areas was considered by Tjostheim (1978) who developed a statistic to measure the degree to which ranked values on two variables occupy positions that are close together in space. The statistic was later generalised by Hubert and Golledge (1982). The statistic is defined:

$$\Lambda = \Sigma_i d(l_F(i), l_G(i))$$

where $l_F(i)$ is the location of rank i on variable F and $l_G(i)$ is the location of rank i on variable G and $d(.,.)$ is a measure of spatial separation. The statistic does not appear to have attracted many applications but measuring the association between the incidence of high levels of respiratory illness and the geographical distribution of air pollution is one area for the reasons discussed above. The statistic is of particular interest because measuring association between variables is complicated in those situations where, because of the nature of the spatial units, exposure levels can be a function of conditions in many units.

This section has reviewed some of the statistical methods for exploring geographically referenced health data and measuring and modelling relationships between health data and socioeconomic and environmental data.

3.3 Constructing an areal system for analysis

The aggregation of health event data to an areal framework involves the loss of some of the original detail although the benefits are that it may then be possible to look at associations between health events and socioeconomic and environmental variables that cannot be studied at the level of individual cases. This may be because of confidentiality reasons or because such data are not available at the level of individuals or because individual level estimates of exposure to risk factors are impossible to compute reliably.

However, one of the consequences of following this route is that the choice of areal aggregation becomes critical. If the aim is to measure the association between subpopulations and socioeconomic and (or) environmental variables then the areas should satisfy a number of important criteria:

- Availability of other information. Socioeconomic data are available in the UK through the census and the smallest areal unit is the enumeration district (ED) containing on average between 125–220 households. The next unit up is the ward which contains on average about 20 EDs. Health event data, which are postcoded, have to be converted to the ED framework (rather than the other way around). This is because postcodes are not census units and because unit postcodes are very small which would result in much suppression of data for confidentiality reasons were the census data to be purchased from the Office of Population Censuses and Surveys (OPCS) in this form.

- Size. Health event data are converted to rates (typically directly or indirectly standardised rates controlling for the population numbers, age and sex composition of the areas). If areas are small, rates are unlikely to be robust in the sense that with small populations the addition or subtraction of small numbers of disease cases in an area can have a large effect on the size of the computed rate. Not only should all areas be of sufficient size it is also arguable that all areas should be of similar size so that rates *between* areas are of equivalent robustness.

- Homogeneity. Linking disease rates to socioeconomic measures (for example, deprivation) will be unsatisfactory even within the terms of ecological analysis if the area contains considerable variability (heterogeneity) with respect to the socioeconomic characteristic. A spatial framework based on wards will suffer from this problem in the case of a variable like material deprivation (Ubido and Ashton, 1993). EDs also contain heterogeneous populations but the heterogeneity is less pronounced than in the case of wards.

Haining *et al.* (1994) constructed a spatial framework for analysing the association between the incidence of colorectal cancer and material deprivation by obtaining the Townsend index of material deprivation for each of the 1159 EDs in the Sheffield Health Authority Metropolitan District from which were produced 48 Townsend Deprivation Regions.

A requirement for this type of analysis is a good regionalisation algorithm. Openshaw and Liang Rao (1994) review some of the algorithms that exist for constructing optimal zoning systems (regionalisations) based on census EDs. Wise *et al.* (1997) review region building algorithms and describe a procedure available in SAGE (see Section 3.5) that is appropriate for meeting the homogeneity and size criteria important in the case of medical geography and which by emphasising speed of operation may be suitable as part of a programme of ESDA.

3.4 Data quality

Quantitative analysis depends critically on data quality. There are several data quality issues that arise in the sorts of analyses identified in the earlier sections.

3.4.1 Health data

Perhaps the main concern here (setting aside issues of clinical diagnosis which is outside the scope of this discussion) is the accuracy with which health events can be assigned to EDs for the purpose of exploring data properties and associations. Address listings can be assigned an exact (to within one metre) National Grid reference using the Ordinance Survey ADDRESS-POINT data but at approximately 12 pence per address this can quickly become expensive. Alternatively, and indeed if only the postcode rather than the full address is available, the Postcode Address File (PAF) can be used which is considerably cheaper but only accurate to within 100 metres. A new directory linking postcodes to EDs has been developed for the 1991 census which is more accurate than PAF. However, both of the cheaper routes run the risk of assigning health events to the wrong ED. In the case of the PAF a 100 metre level of accuracy is clearly a problem in urban areas where EDs are not geographically very large. The 1991 directory has an assignment problem when dealing with postcodes that overlap two or more EDs (Collis et al., 1998).

The implication here is that the locational accuracy of health event data is compromised by any assignment process other than the (expensive) OS ADDRESS-POINT data. When dealing with relatively rare events such misassignment can have a serious impact on estimated rates. Where EDs are grouped into large contiguous areas the risks of inaccurate assignment are reduced.

3.4.2 Census data

Census data are collected every ten years which immediately sets a limit on the accuracy of the recorded counts for periods other than the date of the census itself. (And, of course, there are inherent errors in the census data at the time of collection associated with undercounting for example.) The application of methods described here to non-census years is subject to a declining level of accuracy in the data with inaccuracy being greatest as the time approaches for the next census. One way to reduce the effect is to focus analyses on the period around the date of the census although, of course, this may not always be appropriate. Areas subject to considerable migration or urban redevelopment for example suffer particularly from these effects.

The accuracy of UK census data is also affected by the process of Barnardisation by which counts are randomly altered by 0, ± 1 for reasons of confidentiality. Confidentiality is also the reason for the suppression of data in the case of spatial units with very small populations. In the case of the larger spatial units such as wards this is less of a problem than in the case of small spatial units like EDs. Such inaccuracy in the census data undermines the computation of several important measures described above including the Townsend index of material deprivation, which is based on standardised values of four census variables, and the expected counts on which standardised rate estimates are based.

There is further discussion on the quality of cancer data in terms of diagnosis, statistical coverage and the linkage of the cancer data with residence data in the chapter by Teppo in this volume. An important concern is that when dealing with small regions, small data inaccuracies can have a serious effect on computed rates, including standardised incidence rates, especially in the case of rare diseases. Bayes adjustment is often applied to standardised incidence rates which helps to ensure comparability of rates computed across areas with different observed numbers of cases (Clayton and Kaldor, 1987). Areas with small observed counts have their rates driven towards the regional mean. There is further discussion of the use of Bayes smoothing in Cressie (1992).

3.5 GIS and spatial statistics

There have been a number of purpose written packages built to permit various forms of spatial statistical analysis. Such packages include GAM (Openshaw et al., 1987), INFOMAP (Bailey, 1990), REGARD (Haslett et al., 1991), MANET (Unwin, 1996) and SpaceStat (Anselin 1990). The software has been largely written from scratch, a problem sometimes eased by drawing on toolkits (such as Tcl and Tk) but still necessitating writing software to do things that GIS is already good at.

Approaches that have sought to attach statistical analysis to GIS can be classified into one of two types: loose coupling and close coupling. In the first case linkage is via data files as in the case of linking ARC/INFO and GLIM (Kehris, 1990). In the second case the system is built round one package modified as necessary to allow other features to be incorporated and calling other packages. Examples of this type of coupling include the work of Ding and Fotheringham (1992), Batty and Yichun (1994) who used ARC/INFO as the starting point and Brunsdon and Charlton (1995) and Gatrell and Rowlingson (1994) who used statistical packages (XLisp-Stat and S-Plus respectively) as the starting point. For reviews of work in this area see Goodchild et al. (1992), Haining and Wise (1991) and Haining et al. (1996).

Geographic information systems are potentially of considerable value in implementing the analyses identified in the previous sections. They provide database and mapping facilities and in addition they provide some useful specialist facilities. For example, they facilitate the merging and mapping of data recorded on different spatial frameworks. They also have capabilities that assist region building as described in Section 3.3 where clusters of EDs are merged to form larger areas (Wise et al., 1997). GIS is also potentially of considerable value where the analyst wishes to explore the effects of accessibility. Consider, for example, the uptake of a screening programme like breast cancer which in the case of Sheffield is offered at a single site. To what extent may geographical patterns of uptake reflect problems of public transport access to the screening site from different parts of the city? Suppose there is a proposal to close one or more of several service sites (such as an Accident and Emergency unit). What is the geographical pattern of current usage at each of the existing sites and what might the effects be on travel times for different geographical populations arising from closing any one of the current sites?

Notwithstanding these examples of the use of GIS in this area of research, the current statistical analytical capability of ARC/INFO is still modest. Most of the statistical analysis capabilities described in the preceding sections cannot as yet be implemented within any GIS as far as this author is aware. Currently the author is involved in an ESRC funded project jointly with Stephen Wise and Jingsheng Ma at the Department of Geography, the University of Sheffield, to incorporate spatial statistical analysis capability in ARC/INFO of specific relevance to area-based analyses of health data. This involves drawing on ARC/INFO's powerful database management and map drawing facilities while adding additional modules for other necessary activities that ARC/INFO is either unable to perform or is unsuited to performing. These other activities include graphs and spreadsheets as well as a number of spatial statistical analysis techniques as described above. The prototype version of SAGE (Spatial Analysis in a GIS Environment) has been developed in a client–server architecture with ARC/INFO acting as the server while a program consisting of a number of other visual and non-visual functions complementing those in ARC/INFO acts as a client. The system operates a series of linked windows through which the user can interact rapidly with the data, including highlighting cases in one window (such as a statistical outlier or a set of extreme cases in a box plot or frequency plot) to have them highlighted in other windows including the map window (Haining et al., 1996).

Figures 3.1 to 3.3 illustrate some SAGE sessions. One facility in SAGE allows the user to build regions from small spatial units (for example, enumeration districts) by merging them in ways that allow the user to control for intra regional heterogeneity and inter regional equality of given variables.

Figure 3.1 shows a regionalisation of Sheffield based on 1200 enumeration districts aggregated into about 200 Townsend deprivation regions. The other windows show a histogram of the interquartile range of the enumeration district Townsend scores that comprise each region (homogeneity) and a second histogram of population at risk counts for the set of regions (equality). The regions have been constructed to try to achieve a set of regions with uniform Townsend scores (for the enumeration districts that comprise them) and similar population counts. The histograms show the extent to which these objectives have been realised by the non-hierarchical regionalisation routine that allows the user to specify the number of regions that is required. (For more details of the algorithm see Wise et al., 1997.)

Figure 3.2 shows a session in which a box plot of the 200 incidence rates for colorectal cancer in Sheffield (1979–1983) has been displayed and large (outlier) values from the plot highlighted. These cases are then automatically highlighted in the map and spreadsheet windows. An associated histogram of the Getis–Ord, G_i^*, statistic computed up to and including lag-two adjacency is shown simultaneously and the regions that have been highlighted on the box plot are also highlighted on the histogram. It appears that regions with high incidence rates are not by and large embedded in areas (groups of regions) that have concentrations of high rates since most of the corresponding G_i^* values are not on the right-hand tail of the distribution. Figure 3.3 shows the tail of the histogram of the G_i^* values highlighted. This reveals regions at the centres of concentrations of relatively high incidence rates. The regions are then highlighted on the map, the spreadsheet and the box plot. The response times of the system are good enabling the user to implement interactive ESDA virtually instantaneously as well as confirmatory procedures some of which (such as spatial statistical model fitting) require longer response times (>5 minutes) when the areal system exceeds 500 spatial units.

3.6 Conclusion

The developments reported in this chapter are primarily limited to the linkage of health data with socioeconomic data collected by census units. As noted in the discussion, however, other data may also be relevant to the explanation of the spatio-temporal incidence of disease including environmental data. Environmental data may constitute surfaces rather than areal aggregates so that the spatial index for the observed event is a sample point in two-dimensional space. Examples of this include ground levels of radiation or atmospheric pollution. In order to add environmental data it may be necesary to convert point samples (for example, of air quality) into a surface of air quality so that estimates can be attached to different parts of the city (Collins, 1996). This line of argument suggests that the extension of these methods to this form of health data analysis may require the incorporation of

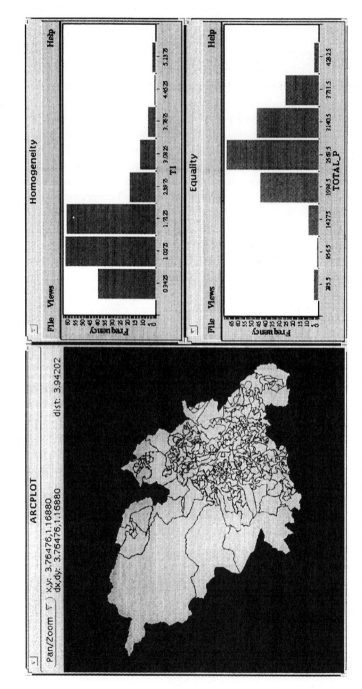

Figure 3.1 Deprivation regions in Sheffield: homogeneity and equal population objectives.

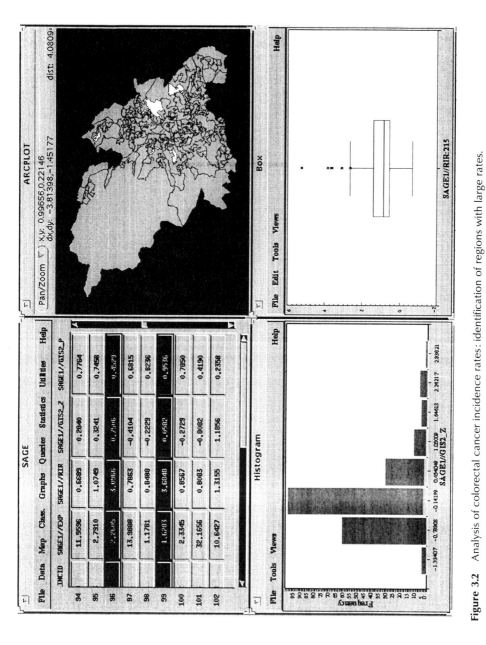

Figure 3.2 Analysis of colorectal cancer incidence rates: identification of regions with large rates.

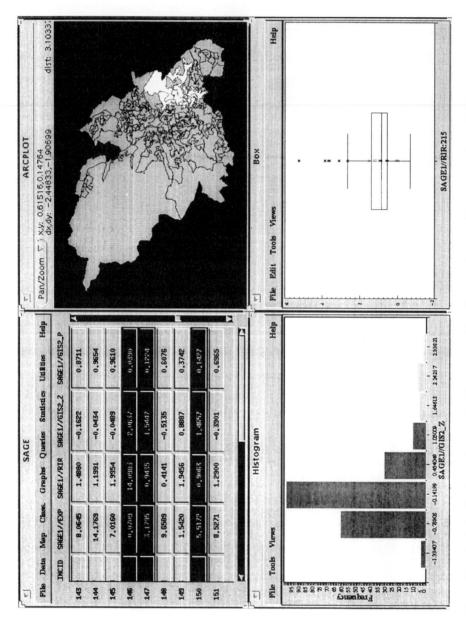

Figure 3.3 Analysis of colorectal cancer incidence rates: identification of areas with concentrations of high rates according to the G_i^* statistic.

spatial interpolation methods, such as kriging, in which point sample readings are interpolated to a larger area together with estimates of sampling error. A number of interpolation methods have been developed (Ripley, 1981). This in turn extends the requirements for GIS if it is to be able to offer the range of analytical capabilities necessary to undertake spatial analysis of health data.

Acknowledgments

The author wishes to acknowledge receipt of ESRC research grant R000234470 which has made the research reported here on SAGE possible. The author also wishes to thank Steve Wise and Jingsheng Ma, who have collaborated on this ESRC project, for many helpful discussions on the subject matter of this chapter.

References

ANSELIN L. (1990) *Space Stat: A Program for the Statistical Analysis of Spatial Data*. Department of Geography, University of California, Santa Barbara.

ANSELIN L. (1995) Local indicators of spatial association – LISA. *Geographical Analysis*, 27(2), 93–115.

BAILEY T. C. (1990) GIS and simple systems for visual interactive spatial analysis. *The Cartographic Journal*, 27, 79–84.

BATTY M. and YICHUN, X. (1994) Urban analysis in a GIS environment: population density modelling using ARC/INFO. Pages 189–220 in Fotheringham S. and Rogerson P. (eds) *Spatial Analysis and GIS*, Taylor and Francis, London.

BRUNSDON C. and CHARLTON M. (1995) A spatial analysis development system using LISP. *Proc. GISRUK '95*, pp. 155–160.

CLAYTON D. and KALDOR J. (1987) Empirical Bayes estimates of age-standardized relative risks for use in disease mapping. *Biometrics*, 43, 671–681.

CLIFF A. D. and ORD J. K. (1981) *Spatial Processes: Models and Applications*. Pion, London.

CLIFFORD P. and RICHARDSON S. (1985) Testing the association between two spatial processes. *Statistics and Decisions*, Suppl. 2, 155–160.

CLIFFORD P., RICHARDSON S. and HEMON D. (1989) Assessing the significance of the correlation between two spatial processes. *Biometrics*, 45, 123–134.

COLLINS S. E. (1996) *A GIS approach to Modelling Small Area Variations in Air Quality*. PhD Thesis, University of Huddersfield.

COLLINS S. E., HAINING R. P., BOWNS I. R., CROFTS D. J., WILLIAM T. S., RIGBY A. and HALL D. (1998) Errors in postcode to enumeration district mapping and their effect on small area analysis of health data. *Journal Public Health Medicine* (forthcoming).

CRESSIE N. A. C. (1991) *Statistics for Spatial Analysis*. Wiley, New York.

CRESSIE N. (1992) Smoothing regional maps using empirical Bayes predictors. *Geographical Analysis*, 24, 75–95.

CRESSIE N. (1994) Towards resistant geostatistics. Pages 21–44 in Verly G. et al. (eds) *Geostatistics for Natural Resources Characterization*, Reidel, Dordrecht.

CRESSIE N. and READ T. R. C. (1989) Spatial data analysis of regional counts. *Biometrical Journal*, 6 699–719.

DE LEPPER M. J., SCHOLTEN H. and STERN R. (1995) *The Added Value of Geographical Information Systems in Public and Environmental Health*. Kluwer, Dordrecht.

DING Y. and FOTHERINGHAM, S. (1992) The integration of spatial analysis and GIS. *Computers, Environment and Urban Systems*, 16, 3–19.

GATRELL A. C. and ROWLINGSON, B. (1994) Spatial point process modelling in a GIS environment. Pages 147–164 in Fotheringham, S. and Rogerson P. (eds) *Spatial Analysis and GIS*, Taylor and Francis, London.

GETIS A. and ORD J. K. 1992. The analysis of spatial association by use of distance statistics. *Geographical Analysis*, **24**, 189–206.

GOODCHILD M. G., HAINING R. P. and WISE S. M. (1992) Integrating geographic information systems and spatial data analysis: problems and possibilities. *International Journal of Geographical Information Systems*, **16**, 407–424.

HAINING R. P. (1990) *Spatial Data Analysis in the Social and Environmental Sciences*. Cambridge University Press, Cambridge.

HAINING R. P. (1991a) Estimation with heteroscedastic and correlated errors: a spatial analysis of intra-urban mortality data. *Papers in Regional Science*, **70**, 223–241.

HAINING R. P. (1991b) Bivariate correlation with spatial data. *Geographical Analysis*, **23**(3), 210–227.

HAINING R. P. (1994a) Diagnostics for regression modeling in spatial econometrics. *Journal of Regional Science*, **34**, 325–341.

HAINING R. P. (1994b) Designing spatial data analysis modules for geographical information systems. Pages 45–64 in Fotheringham S. and Rogerson P. (eds) *Spatial Analysis and GIS*, Taylor and Francis, London.

HAINING R. P. (1996) Designing a health needs GIS with spatial analysis capability. Pages 53–65 in Fischer M., Scholten H. and Unwin D. (eds) *Spatial Analytical Perspectives in GIS*, Taylor and Francis, London.

HAINING R. P. and WISE S. M. (eds) (1991) *GIS and Spatial Data Analysis: Report on the Sheffield Workshop*. ESRC Regional Research Laboratory Initiative. Discussion Paper No. 11.

HAINING R. P., WISE S. M. and BLAKE M. (1994) Constructing regions for small area analysis: material deprivation and colorectal cancer. *Journal of Public Health Medicine*, **16**, 429–438.

HAINING R. P., WISE S. M. and MA J. (1996) Design of a software system for interactive spatial statistical analysis linked to a GIS. *Computational Statistics*, **11**, 449–466.

HASLETT J., BRADLEY R., CRAIG P.S., WILLS G. and UNWIN A. R. (1991) Dynamic graphics for exploring spatial data with application to locating global and local anomalies. *American Statistician*, **45**, 234–242.

HOAGLIN D. C., MOSTELLER F. and TUKEY J. W. (1983) *Understanding Robust and Exploratory Data Analysis*. Wiley, New York.

HUBERT L. J. and GOLLEDGE R. G. (1982) Measuring association between spatially defined variables: Tjostheim's index and some extensions. *Geographical Analysis*, **14**, 273–278.

KEHRIS E. (1990) *A Geographical Modelling Environment Built Around ARC/INFO*. North West Regional Research Laboratory Report 13.

OPENSHAW S., CHARLTON M., WYMER C. and CRAFT A. W. (1987) A Mark 1 geographical analysis machine for the automated analysis of point data sets. *International Journal of Geographical Information Systems*, **1**, 335–358.

OPENSHAW S. and LIANG RAO (1994) Re-engineering 1991 census geography: serial and parallel algorithms for unconstrained zone design. Paper presented to the *Dublin Meeting of the Regional Science Association*, Dublin, 1994.

ORD J. K. and GETIS A. (1995) Local spatial autocorrelation statistics: distributional issues and an application. *Geographical Analysis*, **27**, 286–306.

POCOCK S. J., COOK D. G. and BERESFORD S. A. (1981) Regression of area mortality rates on explanatory variables: what weighting is appropriate? *Applied Statistics*, **30**, 286–296.

RICHARDSON S. (1992) Statistical methods for geographical correlation studies. Pages 181–204 in Elliott P., Cuzick J., English D. and Stern R. (eds) *Geographical and Environmental Epidemiology: Methods for small area studies*. Oxford University Press, Oxford.

RIPLEY B. D. (1981) *Spatial Statistics*. Wiley, Chichester.

TJOSTHEIM D. (1978) A measure of association for spatial variables. *Biometrika*, **65**, 109–114.

UBIDO J. and ASHTON J. (1993) Small area analysis. *Journal of Public Health Medicine*, **15**, 137–143.
UNWIN A., HAWKINS G., HOFMAN H. and SIEGL G. (1996) Interactive graphics for data sets with missing values – MANET. *Journal Computational and Graphical Statistics*, **5**, 113–122 (other information at http://www1.math.uni-augsburg.de/Manet)
WISE S., MA J. and HAINING R. P. (1997) Regionalization tools for the exploratory spatial analysis of health data. Pages 83–100 in Fischer M. and Getis A. (eds) *Recent Developments in Spatial Analysis: Spatial Statistics, Behavioural Modelling and Computational Intelligence*. Springer, Berlin.

CHAPTER FOUR

Statistical Methods for Spatial Epidemiology: Tests for Randomness

MARTIN KULLDORFF

4.1 Introduction

There are many different types of statistical methods useful in spatial epidemiology. Sometimes these are classified according to the nature of the data (area versus point based), or according to the nature of the statistical test (for example, distance based). One can also classify them according to the purpose of the analysis, such as:

- Descriptive (visual) methods for disease maps.
- Map smoothing techniques.
- Spatial regression.
- Spatial–epidemic modelling.
- Small area estimation.
- Tests for spatial randomness.

In this chapter we survey available tests for spatial randomness. This emphasis is a reflection of what Jacquez (page 22) describes as a common situation where 'visually striking thematic maps . . . are not subjected to statistical tests to determine whether an apparent pattern is statistically significant and thus merits explanation'. The same problem can occur with the modelling of spatial epidemics, and we should watch that we do not use increasingly sophisticated techniques to model random spatial noise.

4.1.1 Four types of tests for spatial randomness

For any test of spatial randomness the null-hypothesis is that there is a constant relative risk throughout the study region and that cases occur independently of each other. Different types of problems lead to different types of tests though. In the introductory chapter, Gatrell and Löytönen (page 7) distinguish between:

- Assessing whether there is a raised relative risk around some pre-specified point or line source.

- Detecting whether there is disease clustering.
- Detecting clusters.

To which we add:

- Evaluating cluster alarms.

Tests of the first type are usually called *focused tests* (Besag and Newell, 1991), and they deal with a pre-specified hypothesis concerning the existence of a local cluster in some particular area, a hypothesis that was not generated by the data itself. Methods of the second type will be denoted as *tests for global clustering* in this survey, reflecting their emphasis on testing if a process of clustering is present globally throughout a study region. This occurs when there is an increased risk of seeing cases near other cases, no matter where they occur. Note that there need not be actual cases throughout the area, and the most extreme form of global clustering is when all cases are located close to all other cases, as in Figure 4.1. In this example, out of 40 individuals one case is chosen at random, who then infects their 19 closest neighbours. With tests for global clustering, no inference is made on the location of clusters.

Tests for the detection of clusters are, as focused tests, concerned with local clusters, but are used when we are simultaneously interested in detecting their location and testing their significance. *Evaluating cluster alarms* is maybe the most common type of problem, where someone has looked at the data in an *ad hoc* manner, and found an area with more than its average share of cases. We then want to know if the observed local excess is statistically significant, but we cannot use a focused test since the alarm was generated by the data themselves rather than some external hypothesis.

There is an important difference between a test for global clustering and the other three types of test. With a test for global clustering, the specific location of any case is important no matter where it occurs, since moving it closer or further away from other cases will determine the amount of evidence for global clustering. So, if one case is moved closer to another case, as in Figure 4.2, that would imply a stronger indication of global clustering. With local clusters on the other hand, it is different. If two particular cases happen to be in Baltimore rather than one in Baltimore and the other in Washington, that should not be used as evidence for a local cluster in California at the other end of the country. Likewise, moving one case as indicated in Figure 4.2 does not give additional evidence concerning a potential local cluster in the upper right-hand corner of the map.

While the four different types of tests will tend to have different power for different types of alternative hypotheses, that is not their main distinction. Rather, the

Figure 4.1

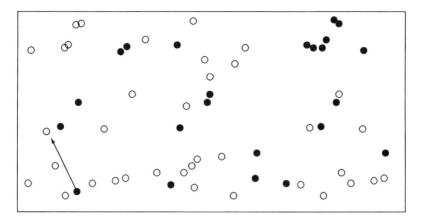

Figure 4.2

main difference is in the type of analysis they perform, where the tests for global clustering make no inference on the location of clusters while the other tests do.

In addition to purely spatial problems, we also survey available methods concerning the distribution of disease in space and time. The most common type of question has been to ask whether there is any space–time interaction, but as with purely spatial data, we might also be interested in global clustering, in tests for the detection of clusters, and in the evaluation of reported space–time cluster alarms. All these are dealt with in Section 4.3.

In Chapter 1, Gatrell and Löytönen (page 7) write that 'if we have addresses of individual cases, it seams perverse, from an analytical point of view, to lose this detail'. Unless explicitly stated to the contrary, all the methods presented in this survey can be applied to either aggregated or individual point-based data, depending on what data and computer resources are available. Moreover, regardless of the level of aggregation, the background population may be represented either by full census population counts, or by a selected set of controls. This is good to keep in mind when reading the original references since some of the tests are described with a particular type of data in mind.

4.1.2 Comparison with spatial regression

In the previous chapter Haining presents a good survey of statistical methods for spatial regression, and it might be worthwhile looking in more detail at the difference between the two types of analysis.

In spatial regression we are interested in inference relating disease risk with various variables that are either geographical in nature, or more commonly, only available on a geographically aggregated level. Examples of such variables are socioeconomic status, air pollution, temperature or smoking levels. A problem with using ordinary regression in a situation like this is that the results could be biased due to unknown spatially related variables affecting the disease risk. In spatial regression, such factors are adjusted for by incorporating purely spatial correlation into the regression model. Tests for spatial randomness do the opposite, doing inference on the purely spatial variation of disease risk after adjusting for various spatially correlated explanatory variables.

4.2 Purely spatial methods

4.2.1 Focused tests

One of the most common inferential problems in the analysis of spatial health data concerns a hypothesis that the risk of a disease is high close to some geographical feature or a collection of geographical features. This could be one or more point sources, like power plants or airports. It could also be a linear source, such as power lines or highways. The same statistical methods can be used in all of these cases.

A common but rough approach is to select a circle of fixed radius around the point source, or equivalently, a corridor of fixed width around the linear source, and compare the frequency of cases inside compared with outside of such a designated buffer. This is not recommendable. The choice of buffer size is arbitrary by default, except in rare circumstances, and the dichotomisation of the data leads to unnecessary loss of information.

There are a number of more suitable statistical methods. How good a statistical test is depends on the power it has against the alternative hypothesis of an increased risk close to the source. That power will depend on the exact shape of the relative risk as a function of the geographical locations. Bithell's linear risk score test (Bithell, 1995) has been shown to be the most powerful test against any simple alternative hypothesis, so if we think that we know the shape of the risk function, at least approximately, then that is the method of choice. It is not clear how robust it is against a misspecification of the alternative though. The Lawson–Waller local score test (Lawson, 1993; Waller et al., 1992) has been shown to be reasonably robust (Waller, 1996; Waller and Lawson, 1995) but it is not known whether it outperforms the Bithell test in this respect. Stone's Poisson maximum test (Stone, 1988) and the focused version of Besag and Newell's test (1991) appears to perform less well (Lumley, 1995; Waller, 1996; Waller and Lawson, 1995). There has been no thorough power study concerning the focused test proposed by Diggle (1990).

Another possible choice is to use binary isotonic regression, which is also known as Stone's MLR test (Lumley, 1995; Stone, 1988). Since it is using a nonparametric risk function, the test should perform reasonably well against any alternative hypothesis as long as the risk decreases as the distance to the source increases. Lumley (1995) has shown that it outperforms Stone's Poisson maximum test, but there has been no study comparing its power to any of the other methods. One useful feature of isotonic regression is that the test result does not depend on whether one uses distance from the source or a measure of exposure for the analysis, as long as the exposure is decreasing with distance from the source.

4.2.2 Global clustering

Tests for global clustering are used when we want to investigate whether there is clustering throughout the study region, but when we are not interested in the specific location of clusters. For example, we might want to know if a particular disease is infectious or not, in which case we would expect cases to be found close to each other no matter where they occur. The most extreme example of this type of clustering arises if we have say one parent case, with all other cases being its closest neighbours (Figure 4.1), but more typical is a situation where groups of cases are spread throughout the map.

Global clustering may occur through two different types of random processes, as indicated by Haining on page 31. It could be a process where initial cases generate other cases with a comparatively higher probability among their closest neighbours, as when a disease is infectious. It could also be that there is a large number of health hazards scattered throughout the region, each creating an increased risk for the disease in a limited surrounding area. Examples of the latter could be gas stations or newly built apartment buildings.

There are a number of tests for global clustering to be used when cases under the null hypothesis are generated by a homogeneous Poisson process, most of which were developed for the botanical sciences including forestry (Diggle, 1983). None of these classical tests can be directly used for spatial health data, since we must now take uneven population density into account in the analysis.

Whittemore *et al.* (1987) have proposed a method entitled a 'test to detect clusters', but according to our classification it is a test for global clustering since it does not determine the location of clusters. It calculates the average distance between all cases of the disease and compares this with the average distance between all individuals, whether a case or not. Clustering is indicated by the former being the lower number.

The test has good power if cases are abundant in the central area of the map, but a power less than the nominal significance level if cases appear with higher probability in peripheral areas. To see this, consider Figure 4.3, where everyone lives in one of nine villages located on a straight line 1 kilometre apart, and with the same size population in each village. Distances within a village are negligible. The average distance between any pair of individuals is then just slightly more than 2 and 26/27 kilometres while the average distance between cases is slightly more than 4 kilometres, and so the Whittemore statistic tells us that there is the opposite of clustering in this data set. In contrast, the remaining tests presented in this section all conclude that there is clustering as we would expect them to do.

The examples used by Tango (1995) illustrate the same phenomena, and the conclusion must be that the Whittemore method is useful to test if there is an increased risk in the central areas of the map, as compared with outlying areas, but it should not be used as a test for global clustering.

Cuzick and Edwards (1990), and independently Alt and Vach (1991), have also proposed a test for global clustering taking the inhomogeneous population density into account. Intuitively it is a very appealing method in the sense that it looks at each case to see if its k nearest neighbours are also cases (rather than controls) to a larger extent than what would be expected under the null hypothesis of a purely random configuration. The variable k can be chosen as fixed, or the test statistic can be taken as a weighted sum over several possible values.

Figure 4.3

Diggle and Chetwynd (1991) have proposed an alternative method, also presented in Diggle (1993), which has become rather popular. They use what is called a K function (Ripley, 1981), defined as:

$$K_{i,j}(s) = \lambda_j^{-1} E[X_{i,j}(s)]$$

where $X_{i,j}(s)$ is the number of (further) type j events within distance s of a randomly chosen type i event, and λ_j is the average number of type j events per unit area. They then propose the test statistic:

$$D(s) = \hat{K}_{1,1}(s) - \hat{K}_{2,2}(s)$$

where cases correspond to events of type 1, and controls to events of type 2. The estimate of the K function, $\hat{K}_{i,i}(s)$, is simply the average number of further type i events within a distance s of a specific type i event, taken over all events i, and then divided by the number of further type i events per unit area, with an additional correction for edge effects. Without the edge correction term, the formula is

$$\hat{K}_{i,i}(s) = |A|/(n_i(n_i-1)) \sum_{k=1}^{n_i} \sum_{m=1}^{n_i} \delta_{km}(s)$$

where n_i is the number of type i events, $|A|$ is the area size, and $\delta_{km}(s)$ is the indicator function equal to one when events k and m are at most a distance s apart and $k \neq m$.

To evaluate the performance of the Diggle–Chetwynd test consider Figure 4.1. One out of the 40 individuals, selected randomly, carries the disease and has infected his or her 19 closest neighbours. It turns out that if the edge of the map is sufficiently far away from all the points, then $\hat{K}_{1,1}(s) = \hat{K}_{2,2}(s)$ with probability one, so that $D(s) = 0$. This is a null result, indicating no clustering whatsoever. In contrast, all the other tests mentioned in this section will conclude that there is significant clustering.

This is clearly a contrived example, and as shown by the authors, there are other alternative hypotheses for which the Diggle–Chetwynd test do have some power. Nevertheless, and in effect, the statistic $D(s)$ measures the relative amount of clustering of two point processes, where clustering is defined as deviations from what would be expected from a homogeneous spatial point process. Such a method is very useful in a situation where we want to compare the clustering tendencies of two different point processes, such as plant species. It is not suitable for testing if a specific type of event shows spatial clustering after adjusting for the inhomogeniety of another spatial point process, which is the kind of problem we encounter in disease clustering. If we are to use the concept of K-functions for disease clustering, a more promising test statistic might be $\hat{K}_{1,1}(s) - \hat{K}_{1,2}(s)$.

Other tests for global clustering have been proposed by Moran (1950), Besag and Newell (1991), Grimson (1991) and Tango (1995), and exclusively for aggregated data, by Walter (1994). Unfortunately, there have not been many comparative power studies evaluating their various strengths and weaknesses under different alternative hypotheses.

A fundamentally different approach is to first construct a density equalised map projection (Merrill et al., 1996), and then apply one of the classical clustering techniques (Diggle, 1983) on the resulting map. The type of projection would influence the result, and since there is no unique projection, it is not clear why such an approach would be advantageous compared with the other methods that incorporate the uneven population density directly into the test statistic. The major advantage of density equalised map projections is most likely rather in the visualisation of the data.

4.2.3 Surveillance/detection of clusters

Most countries have some programme for disease surveillance in place. Typically they monitor how disease rates vary over time as well as their geographical distribution. For the latter, the most common tool is the production of geographical disease atlases, with different rates in different areas indicated by different colours. This is a useful descriptive tool, indicating areas of higher and lower rates, but it does not tell us whether the low and high rates are due to random spatial variation, or if they reflect a varying risk for the disease in different geographical areas.

It would be valuable to complement descriptive disease atlases with some form of statistical inference, indicating what areas have rates that are significantly high or low. To do this, we must take the multiple testing into account, since high and low rates are bound to occur somewhere. A separate p-value for each area measuring that area's 'significance' compared with the rest of the region will give us many 'significant' areas due to chance alone.

An appropriate approach is to use the spatial scan statistic (Kulldorff and Nagarwalla, 1995; Kulldorff, 1995). This method uses a circle of variable size and location to scan the whole map for areas with high or low rates. Using maximum likelihood estimation, a most likely cluster is chosen. The statistical significance of this cluster is then tested taking the multiple comparison into account that resulted from looking for clusters in many different locations and for many different sizes. The method will also detect and evaluate secondary clusters. For a practical application see Hjalmars et al. (1996).

The Cluster Evaluation Permutation Procedure (CEPP), proposed by Turnbull et al. (1990), is a special case of the spatial scan statistic, where the circles have a fixed cluster radius as defined in terms of the population size. It has higher power than the spatial scan statistic if the true cluster size is within approximately 20% of the specified size, but otherwise the power is lower (Kulldorff and Nagarwalla, 1995).

It should be noted that the spatial scan statistic has good power not only for circles but also for compact clusters of other shapes such as squares (Kulldorff and Nagarwalla, 1995). We cannot expect it to have good power for non-compact clusters though, such as a long and narrow area along a river.

It should be noted that there also exist a number of descriptive techniques for the detection of clusters, such as the methods proposed by Openshaw et al. (1987), Besag and Newell (1991), and Rushton and Lolonis (1996). These are useful when creating disease maps illustrating specific areas with high observed rates, but they are unable to do inference on individual clusters in order to determine if the areas have an excess rate that is statistically significant or not.

4.2.4 Evaluation of cluster alarms

During the last decade, there have been frequent occurrences of disease cluster alarms in both Europe and North America. Such alarms might be triggered by the observations of a local doctor or health official, by concerned citizens, or by members of the media. They are often accompanied by considerable worries in the communities affected. Some of the more famous examples are childhood leukaemia in Seascale (Cumbria, NW England), brain cancer in Los Alamos, New Mexico, breast cancer on Long Island, New York and leukaemia in Krümmel, Germany.

That there is a higher observed rate of the disease is usually not in question, since that is normally what caused the alarm in the first place. What we typically want to know is if the higher rate reflects an increased risk for the disease in that area, or whether it is simply a reflection of random spatial variability.

There are two types of tests that sometimes have been, but which should not be, used to evaluate cluster alarms. We should avoid the focused tests described in Section 4.2.1 since they will give us an erroneous p-value due to pre-selection bias. Indeed, if we were to (i) look at any disease in any geographical region, (ii) find the sub-area with the highest rate within that region, and then (iii) apply a focused test to that location, then we would always get a very small 'p-value' and in most cases a 'significant' one.

We should also avoid using any of the global clustering tests described in Section 4.2.2. While they will have the correct p-value for what they are designed for, they may give a significant result no matter how few cases there are in the area of interest due to the relative location of cases in a completely different part of the map.

There are two options for how statistically to evaluate cluster alarms in a proper fashion.

The first possible approach is to forget about past and present data, and instead monitor future cases as they occur in the area of the alarm. This is a confirmatory type of analysis, and it avoids the pre-selection bias since the analysis is based on cases diagnosed after the alarm occurred. The most attractive approach is to use a sequential procedure with a stopping rule for when to declare the cluster confirmed or rejected, as proposed by Chen et al. (1993). One drawback of the approach though is that it may take a long time to get the results since we have to wait for new cases to occur. This is especially true if the area of the alarm is small, or if we are dealing with a rare disease.

Another approach is to expand the study in space rather than in time. If we collect data for a larger region in which the cluster alarm is located, such as the whole country, we could then apply the spatial scan statistic to that expanded region and see if there is a significant cluster where we would expect it to be based on the alarm. This will give us a measure of how unlikely it is to encounter the observed excess of the cluster alarm in a larger area of our choice. Many cluster alarms could be quickly dismissed as a random occurrence were this technique to be used.

4.3 Space–time methods

4.3.1 Evaluating cluster alarms

In many cases a cluster alarm is not only related to a specific area, but it is also claimed to be present during a limited time period. The arguments for not using a focused space–time test, or a space–time test for global clustering, is the same as for purely spatial cluster alarms. In addition, we cannot use a sequential confirmation procedure either. If the increased risk causing a cluster is limited in time, then future observations will tell us nothing about it.

What is available is the space–time scan statistic (Kulldorff et al., 1998), a generalisation of the spatial scan statistic. Instead of a two-dimensional circle it uses a three dimensional cylinder of variable size, where the circular base represents a particular geographical area and where the height represents a number of consecutive years or months. The cylinder is then moved through space and time in order to detect the most likely cluster, and a p-value is calculated for that cluster, taking the multiple comparison into account.

4.3.2 Surveillance/detection of clusters

There are two possible reasons why we might want to detect the location of clusters that are limited in time in addition to space. Firstly, we could be interested in the etiology of a disease, so that past clusters are just as interesting as current ones, since either may provide valuable clues to what caused them. Secondly, we might want quickly to detect any newly occurring local health hazards in order to implement whatever public health measures are needed. If we simply did a purely spatial analysis covering say a 20-year period, then data from the last couple of years would be diluted through the average taken over the whole time period.

Equivalent to the purely spatial problem, the natural choice for detecting space–time clusters is the space–time scan statistic (Kulldorff et al., 1998), using a cylinder instead of a circle as described in the previous section.

4.3.3 Space–time interaction

Suppose we want to know if a disease is infectious or not. If it is, then we might expect that cases that are close in space are also close in time. This can be investigated through a test for space–time interaction. The most commonly used test for this type of problem is the Knox (1964) test. Other methods have been proposed by Mantel (1967), Diggle et al. (1995) and Jacquez (1996a).

With the Knox test one needs to specify two threshold values, determining how near two cases must be in time and in space respectively in order to be considered close to each other. Diggle et al. (1995) propose a summary statistic with different weights given to a finite set of threshold values. Mantel (1967) replaces the threshold with a general function, in effect assigning weights to each possible threshold value.

The Jacquez (1996a) test does not define the threshold in terms of actual distance, but rather in terms of the k nearest neighbours. This test is invariant under a shift in the population density, which is intuitively sensible, making it equally sensitive in rural and urban areas. The test statistic is either based on a single value of k, or on a summary statistic utilising several different values.

In a power comparison, Jacquez (1996a) showed that his test, followed by Mantel's and then Knox's, performed the best. The test by Diggle et al. (1995) was not included in the comparison, but might by its nature fall somewhere between the Knox and the Mantel tests. All four tests have the very useful feature that they do not require any knowledge about the background population and its geographical distribution, nor do they require any controls. Along with this feature comes a serious limitation though. While the tests are valid for any inhomogeneous geographic population distribution, as well as for any temporal change in the population size, they are all biased if the temporal change is of a different magnitude in different areas of the map. To make things worse, the bias is liberal, so that we will reject the null-hypothesis of no clustering more often than we should according to the nominal significance level.

This bias was first pointed out by Klauber and Mustacchi (1970), who proposed to divide the time period into several sub-intervals, doing the test separately on each one and then combining the results. This procedure reduces the bias but it does not eliminate it. At the same time, power is lost due to information being lost when neighbouring cases fall on different sides of the time cuts.

Gilman and Knox (1995) apply the Knox test to childhood leukaemia in central Britain during 1953–1980. Out of 45 different threshold combinations, there were 17 'significant' results using the original test. When implementing the Klauber–Mustacchi bias reduction, the number of 'significant' results drops to 3, and it is impossible to say how many of these are due to the remaining bias.

Any p-value based on these methods should be treated with great care, and especially when coming from data sets spanning more than a couple of years. Kulldorff and Hjalmars (1998) have shown that the bias can be considerable for certain data sets, and it is then better to use the unbiased variant of the Knox test that they propose, which adjusts for any unequal shifts in the population distribution.

It is important to realise that tests for space–time interaction are designed to detect space–time clustering above and beyond any purely spatial or purely temporal clustering. As an illustration, consider Figure 4.4. There is clearly temporal clustering, since all cases occur in the third and fourth years, and there is clearly spatial clustering, since all cases occur in the upper part of the map, but there is no space–time interaction. Cases that are close in space are not more likely to also be close in time, and vice versa. All four tests will give a null result for this data set, which can be seen by remembering that the methods only use information regarding the cases, completely ignoring years 1 and 2 as well as the lower part of the map.

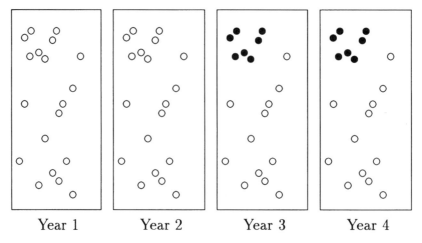

Figure 4.4

4.3.4 Global space–time clustering

There is a natural approach to overcome the bias problem of the space–time interaction tests and at the same time test for space–time clustering even when there is no space–time interaction, the difference being illustrated in Figure 4.4. This is to extend one of the tests for global clustering of purely spatial data by adding a third dimension representing time. Ranta *et al.* (1996) have done exactly this, for the Whittemore *et al.* test (1987), and it could also be done, very easily, for Cuzick and Edwards' test (1990).

4.4 Discussion

As shown, there are many different tests for spatial randomness, and for any practical application it is important to choose wisely between them. This selection process could be divided into three steps. Firstly, we need to distinguish between the different types of problems, and choose our test accordingly. It is hoped this survey will be of assistance in doing that. Secondly, we should try and eliminate from consideration any test for which there are examples of obvious clustering which it cannot detect. For this reason, we have presented a number of especially constructed examples to that effect. Thirdly, of the remaining tests we should try and select whichever has the highest power for the types of alternatives that we think are most likely. When power comparisons do exist, we have mentioned their basic results but we strongly encourage the reader to look in the original references for details. In general though, there are very few power comparisons and a lot of work needs to be done in this area before any conclusive recommendation can be made.

4.4.1 Detecting areas with low rates

In the introductory chapter of this book, Gatrell and Löytönen (page 4) state that: 'We tend to dwell usually on areas of high disease risk, often neglecting to note that areas with low rates may also give clues about disease causation'. This is an important point, and it should be noted that many of the methods mentioned in this survey can be used to study local areas of low instead of high rates by simply renaming each control to a case and vice versa.

4.4.2 Uncertain locations of cases and controls

In geographical analysis, we never have access to the 'exact' geographical locations of cases and controls. This is for two reasons: (i) we often do not know the exact coordinates of residence (birth, workplace, etc.), but only the administrative area in which people live, and (ii) people move around as they go about their daily life (Jacquez, 1994, 1996b). An important question is how this uncertainty of location affects the various tests for spatial randomness.

To understand this, we must look at the principles behind statistical hypothesis testing. The most basic feature is the significance level (type I error) describing the probability of falsely rejecting a true null-hypothesis. For all the tests, the correct significance level is obtained for any uneven population distribution, whatever that distribution is. It does not matter if it reflects the 'true coordinates', if it reflects some approximation, or even if the coordinates are just randomly selected. The significance level will still be correct, and there is no need to make any adjustments for this reason.

This leads us to the second aspect of statistical hypothesis testing, that of power, describing the probability of not rejecting the null-hypothesis for a given true alternative hypothesis (type II error). Mis-specification in the geographical coordinates will typically reduce the power of the tests, and if coordinates are chosen at random, the power will be equal to the significance level, making the test completely worthless.

Hence, the question of mis-specified locations of cases and controls is in essence a question of power. One way to view mis-specified coordinates is as a mis-specified alternative hypothesis. For any 'mis-specification' of the coordinates, there is always a corresponding 'mis-specification' of the alternative hypothesis, leading to no loss in power when they are implemented simultaneously. Hence, a good way to evaluate whether a particular test is robust toward uncertainties in the geographical coordinates, is to see if it is robust to a wide set of slightly modified alternative hypotheses.

4.4.3 Edge effects

In the study of spatial point patterns generated by a homogeneous Poisson process, one typically observes the process only in a limited region. In effect, we then have an inhomogeneous Poisson point process, with a positive constant intensity inside the region and zero intensity outside. In order to still use the powerful theory of homogeneous Poisson processes, it is necessary to take the 'inhomogeniety' into account by adjusting for edge effects (Diggle, 1983), or else the nominal significance level would be incorrect.

For the statistical tests described in this survey the situation is very different. They are by nature designed to adjust for the underlying inhomogeneous population distribution. This adjustment reflects areas where there is no population, and in an identical way, areas outside the study region for which there are no data. Therefore, all tests presented in this chapter are valid and unbiased, and do not need to be further adjusted for edge effects. An 'edge adjusted' test is simply a slightly different test with a slightly increased or decreased power depending on the alternative hypothesis at hand.

References

ALT K. W. and VACH W. (1991) The reconstruction of 'genetic kinship' in prehistoric burial complexes – problems and statistics. In Bock H. H. and Ihm P. (eds) *Classification, Data Analysis, and Knowledge Organization*, Springer Verlag, Berlin.

BESAG J. and NEWELL J. (1991) The detection of clusters in rare diseases. *Journal of the Royal Statistical Society* Series A, **15**, 4143–4155.

BITHELL J. F. (1995) The choice of test for detecting raised disease risk near a point source. *Statistics in Medicine*, **14**, 2309–2322.

CHEN R., CONNELLY R. R. and MANTEL N. (1993) Analyzing post alarm data in a monitoring system, in order to accept or reject the alarm. *Statistics in Medicine*, **12**, 1807–1812.

CUZICK J. and EDWARDS R. (1990) Spatial clustering for inhomogeneous populations. *Journal of the Royal Statistical Society* Series B, **52**, 73–104.

DIGGLE P. J. (1983) *Statistical Analysis of Spatial Point Patterns*. Academic Press, London.

DIGGLE P. J. (1990) A point process modelling approach to raised incidence of a rare phenomenon in the vicinity of a pre-specified point. *Journal of the Royal Statistical Society* Series A, **156**, 349–362.

DIGGLE P. J. (1993) Point process modelling in environmental epidemiology. In Barnett V. and Turkman K. F. (eds) *Statistics for the Environment*, John Wiley, Chichester.

DIGGLE P. J. and CHETWYND A. D. (1991) Second-order analysis of spatial clustering for inhomogeneous populations. *Biometrics*, **47**, 1155–1163.

DIGGLE P. J., CHETWYND A. D., HÄGGKVIST R. and MORRIS S. (1995) Second-order analysis of space-time clustering. *Statistical Methods in Medical Research*, **4**, 124–136.

GILMAN E. A. and KNOX G. (1995) Childhood cancer: space-time distribution in Britain. *Journal of Epidemiology and Community Health*, **49**, 158–163.

GRIMSON R. C. (1991) A versatile test for clustering and a proximity analysis of neurons. *Methods of Information in Medicine*, **30**, 299–303.

HJALMARS U., KULLDORFF M., GUSTAFSSON G. and NAGARWALLA N. (1996) Childhood leukemia in Sweden: Using GIS and a spatial scan statistic for cluster detection. *Statistics in Medicine*, **15**, 707–715.

JACQUEZ G. M. (1994) Cuzick and Edwards' test when exact locations are unknown. *American Journal of Epidemiology*, **140**, 58–64.

JACQUEZ G. M. (1996a) A k-nearest neighbor test for space-time interaction. *Statistics in Medicine*, **15**, 1935–1949.

JACQUEZ G. M. (1996b) Disease cluster statistics for imprecise space-time locations. *Statistics in Medicine*, **15**, 873–885.

KLAUBER M. R. and MUSTACCHI P. (1970) Space-time clustering of childhood leukemia in San Francisco. *Cancer Research*, **30**, 1969–1973.

KNOX G. (1964) The detection of space-time interactions. *Applied Statistics*, **13**, 25–29.

KULLDORFF M. (1997) A spatial scan statistic. *Communications in Statistics: Theory and Methods*, **26**, 1481–1496.

KULLDORFF M., ATHAS W. F., FEUER E. J., MILLER B. A. and KEY C. R. (1996b) Evaluating cluster alarms: a space-time scan statistic and brain cancer in Los Alamos. *American Journal of Public Health* (in press).

KULLDORFF M. and HJALMARS U. (1998) The Knox methods and other tests for space-time interaction. *Biometrics*, to appear.

KULLDORFF M. and NAGARWALLA N. (1995) Spatial disease clusters: detection and inference. *Statistics in Medicine*, **14**, 799–810.

LAWSON A. B. (1993) On the analysis of mortality events associated with a prespecified fixed point. *Journal of the Royal Statistical Society* Series A, **156**, 363–377.

LUMLEY T. (1995) Efficient execution of Stone's likelihood ratio test for disease clustering. *Computational Statistics and Data Analysis*, **20**, 499–510.

MANTEL N. (1967) The detection of disease clustering and a generalized regression approach. *Cancer Research*, **27**, 201–220.

MERRILL D. W., SELVIN S., CLOSE E. R. and HOLMES H. H. (1996) Use of density equalizing map projections (DEMP) in the analysis of childhood cancer in four California counties. *Statistics in Medicine*, **15**, 1837–1848.

MORAN P. A. P. (1950) Notes on continuous stochastic phenomena. *Biometrika*, **37**, 17–23.

OPENSHAW S., CHARLTON M., WYMER C. and CRAFT A. W. (1987) A Mark 1 geographical analysis machine for the automated analysis of point data sets. *International Journal of Geographical Information Systems*, **1**, 335–358.

RANTA J., PITKÄNIEMI J., KARVONEN M. et al. (1996) Detection of overall space-time clustering in non-uniformly distributed population. *Statistics in Medicine*, **15**, 2561–2572.

RIPLEY B. D. (1981) *Spatial Statistics*. Wiley, Chichester.

RUSHTON G. and LOLONIS P. (1996) Exploratory spatial analysis of birth defect rates in an urban population. *Statistics in Medicine*, **7**, 717–726.

STONE R. A. (1988) Investigation of excess environmental risk around putative sources: statistical problems and a proposed test. *Statistics in Medicine*, **7**, 649–660.

TANGO T. (1995) A class of tests for detecting 'general' and 'focused' clustering of rare diseases. *Statistics in Medicine*, **14**, 2323–2334.

TURNBULL B. W., IWANO E. J., BURNETT W. S., HOWE H. L. and CLARK L. C. (1990) Monitoring for clusters of disease: application to Leukemia incidence in upstate New York. *American Journal of Epidemiology*, **132**, S136–S143.

WALLER L. A. (1996) Statistical power and design of focused clustering studies. *Statistics in Medicine*, **15**, 765–782.

WALLER L. A. and LAWSON A. B. (1995) The power of focused tests to detect disease clustering. *Statistics in Medicine*, **14**, 2291–2308.

WALLER L. A., TURNBULL B. W., CLARK L. C. and NASCA P. (1992) Chronic disease surveillance and testing of clustering of disease and exposure: application to leukaemia incidence and TCE-contaminated dumpsites in upstate New York. *Environmetrics*, **3**, 281–300.

WALTER S. D. (1994) A simple test for spatial pattern in regional health data. *Statistics in Medicine*, **13**, 1037–1044.

WHITTEMORE A. S., FRIEND N., BROWN B. W. and HOLLY E. A. (1987) A test to detect clusters of disease. *Biometrika*, **74**, 631–635.

CHAPTER FIVE

Improving the Geographic Basis of Health Surveillance using GIS

GERARD RUSHTON

5.1 Introduction

Gatrell and Löytönen (1996) challenge a basic premise of this book by asserting that 'spatial analysis of epidemiological data has proceeded quite comfortably without GIS!' Yet, many investigators of disease patterns in the public health community in the US are keen to make GIS-based investigations a critical part of their activity. There are two reasons for this apparent contradiction. The first is a difference in purpose between the classical epidemiologist and the public health investigator. The latter, working closely with the community, is far less interested in answering the question posed: 'Can one conclude that disease x occurs in clusters in this region?' than in answering the less clear, but often more pressing question posed by the public: 'Show me the geographical pattern of the disease rate of x in this region and tell me why I should be confident there is no reason to be concerned about the high rate here?' The public is far more interested in the latter than the former, especially in the early stages of any disease investigation as a preparatory step in prioritizing areas that may warrant field investigation. The difference between the two questions is, of course, the difference between confirmatory and exploratory analysis of disease distributions (Bailey and Gatrell, 1995). Geographical information systems, as currently constituted, are useful for exploratory spatial analysis (Gatrell and Bailey, 1996), and less useful for confirmatory analysis, though it is clearly possible to couple methods of confirmatory analysis to a GIS (Diggle and Rowlingson, 1994; Rowlingson and Diggle, 1993; Kingham *et al.*, 1995; Openshaw, 1994).

The second reason that GIS is little used in epidemiological analysis is that 'the spatial analysis of epidemiological data' has generally meant the analysis of disease incidence data as measured and recorded for small areas. Central to the use of GIS is control over the spatial referencing of health and other information – including environmental, demographic, and health resources – to which the disease data are to be related. The ability to search for potential spatial associations between these four conditioners of health status in any area, especially when they are geo-referenced differently, is what attracts public health investigators and the public to GIS as a tool to assist them in examining health and disease patterns in local areas.

So long as the spatial analysis of epidemiological data meant the analysis of data as constituted in common, discrete, spatial units, it could proceed without GIS. In all of the methods we discuss in this chapter, however, we assume that point-based health data are superior to area-based data. Although a recent paper questions this assumption (Oden et al., 1996), common logic is a sufficient basis for concluding that health data that are independently geo-coded are more useful than data that have been placed in arbitrary spatial units. This is especially the case when other information to which a researcher expects health data may be related, is geo-coded to other spatial units, making necessary the co-registration of the health and other information. GIS brings efficiencies and qualitative improvements to spatial analysis.

Missing from most evaluations of the potential contribution of GIS in public health is the identification of opportunities to do new things with GIS that would have been difficult if not impossible to do without it.

GIS provides opportunities to pursue two types of spatial analysis that cannot be done well without it:

- Determine, by exploratory spatial analysis of independently geo-coded health events, areas of high disease incidence that can be labelled as 'statistically worthy of further investigation' (Neutra et al., 1992).
- Determine spatial relationships between disease incidences and other information that is geo-referenced differently from the disease data.

Both types of analysis are best pursued in an interactive, exploratory, computing environment. Although all exploratory analyses are interactive, not all interactive analyses are exploratory. Exploratory analyses investigate alternative understandings of the pattern of disease in any region with the aim of selecting more appropriate subsequent analyses that might resolve the ambiguities that typically arise in the early stages of investigations. Interactive analyses move in a dynamic way between graphical, statistical and other methods of viewing information (Gatrell and Bailey, 1996, p. 844).

In a number of interactions with public health investigators in the US, we have tried to identify specific ways in which value can be added to their investigations by GIS. We have developed and made available on a CD-ROM, two exploratory spatial analysis modules to fill two critical gaps in typical investigations of disease patterns that GIS appears to be ill-equipped to fill at the present time. The CD-ROM (Rushton et al., 1996) is organised to meet the need for educating and training public health investigators in these materials and methods, as well as providing software to complement typical GIS functions that will enable them to analyse their own data using GIS.

The purpose of this chapter is to discuss methods of analysis that describe local patterns of disease incidence and address the issue of the statistical significance of localised rates while maintaining the goal of exploratory spatial analysis. We develop a spatially continuous measure of significance as a key support tool in understanding the pattern of high rates of infant mortality in localised areas of a city. Our aspiration level, as Schweder and Spjotvoll (1982) noted for their method, is to make 'informal inference' about specific areas from which decision-makers can weigh the evidence and take actions they consider appropriate for the case before them. We let the study area serve as its own control by providing a method for comparing the spatial pattern of a disease within the area with other patterns expected there based on Monte Carlo simulations.

5.2 Geographic encoding of health events

In traditional approaches, health events are recorded for small geographic areas. The health characteristics of populations of larger areas are determined by spatial aggregation of the smaller areas. This method of geographic encoding of health events continues in most countries, including the United States. Such health events include vital statistics, disease events in registries and ambulatory and hospital utilisation records. The result is that disease clusters cannot be investigated unless their size and boundaries coincide at least roughly with the spatial units for which the data have been encoded (Waller, 1996). This limitation has been widely recognised ever since the pioneering work of Choynowski (1959) although efforts to deal with it have usually faltered over the absence of alternative methods for spatially encoding health data. With GIS and the increasing availability over much of the developed world of digital spatial databases that either contain household locational identifiers or provide location information to which addresses on individual records can be matched, the placing of household records into pre-defined regions is no longer a rational or even efficient method of spatial data encoding. The practice of coding information on health records to small areas effectively strips spatial information from the health record and substitutes a grosser level of spatial accuracy thereby reducing its value in subsequent spatial analyses (Jacquez, 1996; Jacquez and Waller, 1996; Lovett et al., 1990; Twigg, 1990). It is costly, inefficient and ineffective. California, for example, is only one of many US states that continues to employ 'look-up' tables to locate vital statistics records and encode census tract IDs and five-digit ZIP codes as locational identifiers. This pre-GIS methodology is now fully ensconced in a computerised system accessible to public health professionals there (Strassburg and Williams, 1995). One of the central issues in improving the analysis of disease distributions, health facility utilisation patterns and the optimum geographical location of health resources, therefore, is the basic geographic encoding of health records and the circumstances surrounding accessibility to such records. Clearly, privacy and confidentiality issues are important and the detailed geographic information about a health record needs the same protection as other sensitive information on the record (Institute of Medicine, 1994), but the current practice of stripping away geographic information to protect the privacy of the record is a practice that throws out the baby with the bathwater. Developing new methods for protecting confidentiality of health records while conserving their geographic accuracy should have high priority in future GIS-based research in health. Duncan and Pearson (1991) discuss disclosure limiting procedures for protecting the privacy of individual records in databases.

In the US, 'TIGER' files are increasingly used for automatically assigning geographic codes to health records (US Bureau of the Census, 1992). Because these files are in the public domain, access to them is widespread and inexpensive. Covering the entire country, these files contain the approximate latitude and longitude coordinates for each encoded road link and the ranges of the addresses contained on the left and right sides of each link. Using computer software, any address can be matched to these address ranges and, by interpolation, an estimate can be made of the latitude and longitude of each address. Because the boundaries of political and census administrative areas are also encoded in TIGER records, any address-matched record can be assigned, using the 'point-in-polygon' features of GIS software, to any of these areas. Further discussion and references to TIGER files can be found in Rizzardi et al. (1993); Rushton and Lolonis (1996), p.718; and Rushton et al. (1996).

Adding random digits to the latitude/longitude coordinates of address-matched records is only one of several methods available to limit the disclosure and protect the privacy of such health records though the interesting scientific questions of how such investigator-induced spatial errors may affect the validity of subsequent analyses remains to be investigated.

5.3 Spatial aggregation of disease data and geographic scales of analysis

Most traditional analyses of disease patterns examine disease rates at a given level of spatial resolution defined by spatial entities developed for other purposes. These analyses portray the pattern of disease rates on a choropleth map and show rates with different statistical reliability in different areas (Langford, 1994). Such maps do not control spatial scale since the size of the areas on such maps is variable (Moulton et al., 1994). The question of the differential statistical reliability of information shown (Kennedy-Kalafatis, 1995; Marshall, 1991) has received a lot of attention. The statistical significance of these rates is often determined by comparing each rate with an expected Poisson distribution of events to establish confidence limits (Dobson et al., 1991). The results of such analyses, however, are insensitive to the relative location of the areas. If, for example, the areas were aligned along a linear route from south to north, it is possible that all areas in the north may have positive differences and all in the south may have negative differences. Such patterns would not be expected to arise by chance even though analyses of the distribution of differences might show a pattern not significantly different from zero. Compare Reynolds et al. (1996) for an example of such an analysis where the spatial patterns of the differences are not examined with Hjalmars et al. (1996), where, by combining data for contiguous areas in a spatial scan statistic that encompasses a fixed population base, the spatial patterns of the differences are examined. Stone (1988) also recognises the importance of investigating the spatial pattern of the differences between observed and expected disease rates.

Analyses of disease distributions at multiple spatial scales are rare (Schneider et al., 1993; Waller and Turnbull, 1993). Disease data in the US are usually examined at the level of the census tract, the county, or groups of counties. It is well-known that patterns found at one geographic scale of analysis may be different from the patterns found at another. With GIS, it is relatively easy to analyze the spatial distribution for differences. Such analyses are readily accomplished although all of the issues in interpreting such disease maps remain (Moulton et al., 1994). It is likely that these issues will rise in importance as the number of disease maps at different spatial scales that are produced grows with the widespread adoption of GIS.

If individual-level, address-matched data, at a fine scale of geographic resolution were to become widely available, however, then the conventional methods for visualising pattern, described above, should be abandoned. From the viewpoint of an analyst who controls individual, address-matched disease data, the choropleth map is a spatially filtered map using a non-overlapping, variable-size, spatial filter with filter shapes selected from available political or administrative regions. The choice to process and visualise disease data in this way cannot be rationally defended. Its continued use should be accepted only as a consequence of restricted data availability. Why data should be made available for such odd-shaped and different sized regions can be best explained by one of three factors:

- Data for such areas easily can be encoded from the information provided.
- Information is often requested for such areas.
- Recording health information for areas reduces the risk of disclosure and protects the privacy of individuals.

Developments in GIS lead us to question the continued validity of these factors. It is now as easy in the US to code a health record as a latitude/longitude coordinate as to code it for areas such as census tracts. People request information for political or administrative areas because they are familiar with these areas and use them to convey the spatial limits of their interest. Finally, there are other approaches to limiting disclosure that preserve, better than spatial aggregation methods, the spatial characteristics of the data.

To meet the purpose of exploratory spatial analysis, health data are better examined by methods that assume that disease rates are spatially continuous. Since disease rates change over geographical space, the place where change occurs should not be an artifact of the analysis method used nor of the fundamental data unit examined. Accordingly, both the definition of the problem as well as the increasing availability of individually localised health data lead to the conclusion that flexible spatial filtering of information is the preferred method of exploratory spatial analysis of disease distributions (Bithell, 1990; Bailey and Gatrell, 1995).

5.4 Spatial filtering of disease distributions

The principal reason to filter disease data spatially is to examine the spatial pattern of disease at different levels of spatial resolution and to compute disease rates that are not dependent on the specific boundaries of the areas used in spatially aggregated data. Just as a traveller flying at an altitude of 10 000 metres sees a pattern of land use quite different from the pattern seen at an altitude of 1000 metres, so different spatial patterns of disease are revealed by analyses that use large spatial filters with analyses that use smaller filters. Different diseases have patterns that are interesting at different spatial scales. The optimum sized spatial filter is the size that reveals the most interesting pattern. The GIS should be able to filter the disease cases and the people at risk, for areas that in size and shape can be controlled by the user (Diggle, 1991; Turnbull et al., 1990; Weinstock, 1981).

One simple, spatial filter method is the 'sliding window' method of Weinstock (1981) for estimating the disease rate at each node of a regular grid superimposed over the area of interest. A more general and interesting discussion of spatial filters is found in Brillinger (1994). For the area surrounding each node on a grid, the number of disease incidences and the population at risk are summed and the ratio is the disease rate for the grid location. The results are sensitive to a number of critical

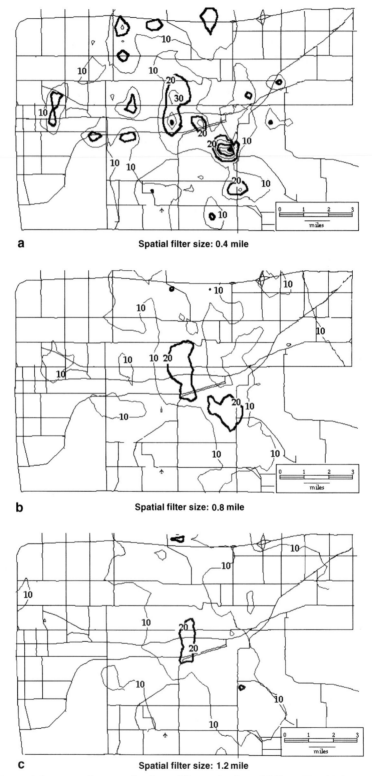

Figure 5.1a–c Infant mortality rates for three different sized spatial filters, Des Moines, Iowa, 1989–1992.

choices. For resolution above a given spatial scale, making the grid finer – that is, increasing the number of grid points, will not change the pattern. Changing the size of the spatial filter – the circle centred on each grid point over which the disease rate is measured – will affect the pattern, in the sense that differences in rates that typically occur within the size of the filter will be averaged or smoothed and some of the variability in the geographic pattern will disappear. A key question in exploratory spatial analysis is when the variability that is averaged by the filter is real and important – and therefore should not be removed – and when it is occurring by chance and therefore is noise that should be removed. This is a difficult question, for to answer it correctly often requires knowledge about the pattern that the analysis is designed to discover. The reference distributions in the significance tests described below are designed to answer this question. Since there is no universal choice of the radius of the spatial filter for all data, it is advisable to carry out the moving window procedure for a few different representative values of r – the radius of the filter (Silverman, 1978).

Illustrations of the application of this logic are shown in Figures 5.1a–5.1c where individual, address-matched records of infant deaths in Des Moines, Iowa, were converted to latitude/longitude coordinates as were the approximately 20 000 births in the area for the same 1989–1992 period. A grid with intersections at approximately 0.4 mile intervals provided the nodes from which infant deaths and births were counted for overlapping areas of three different filter sizes, corresponding to the three maps in Figure 5.1. Infant mortality rates were computed for each grid intersection and a contouring program in the GIS was used to draw isopleth curves describing the surface of the disease distribution. Computational issues of relating these analyses to a typical GIS environment are described in Section 5.6.

5.5 Spatial significance of disease rates

Because the interest here is with exploratory spatial analysis, we are less interested in disease cluster investigation methods that test for the presence or absence of clusters in general (Cuzick and Edwards, 1990; Diggle and Chetwynd, 1991) and more interested in methods that indicate the likelihood that clusters exist at particular locations – as Besag and Newell (1991) describe it – the detection of clusters. A middle case is where tests are developed to detect clusters at particular geographic scales – without attempting to locate the clusters in question (Kingham et al., 1995, p. 813).

To identify specific localities as possible disease clusters, we adopt and develop further the methods of Openshaw et al. (1987, 1988a, 1988b) in their geographical analysis machine. As with Openshaw et al. we lay down a closely spaced grid lattice over the study area and count disease incidences and numbers of people at risk for overlapping circles centred on each grid point. Unlike Openshaw et al., Fotheringham and Zhan (1996), Kulldorff and Nagarwalla (1995) and others who have adopted variants of this approach, we keep the size of the circle constant in any analysis. In contrast, Turnbull et al. (1990), and Hjalmars et al. (1996), hold constant the population size of each cluster investigated. Their approach leads to a more acceptable and more easily computed statistical basis for tests of significance but, unfortunately, varies the spatial scale at which clustering is defined and hence,

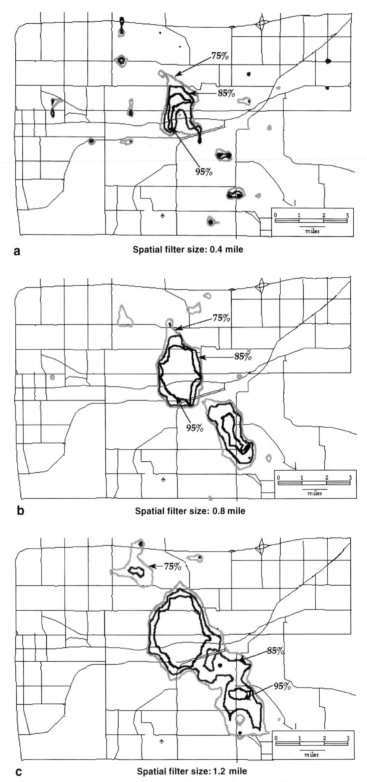

Figure 5.2a–c Significance levels for infant mortality rates for three different sized spatial filters, Des Moines, Iowa, 1989–1992.

detected. Here we test the significance of clusters at the given geographic scale as determined by the size of the circle (spatial filter) used and compute the parameters of a reference distribution separately for each grid point. We conduct a Monte Carlo simulation of infant deaths in which each birth has an equal and identical probability of becoming a death. The method can accept as input any probabilities for the event, thus incorporating any prior information about the characteristics that affect the likelihood of a given case becoming diseased. For each simulation, we count the number of simulated cases in the same circle. From 1000 such simulations, the proportion of the simulated disease rates at each grid point that were less than the observed disease rate at the point is established – the reference distribution – and a spatial interpolation of these proportions is the map of statistical significance that the observed rate was larger than expected, given the null hypothesis (Figures 5.2a–5.2c). A more detailed description of the method as applied to cases of birth defects in the same area is found in Rushton and Lolonis (1996).

Since the number of deaths in any filter area is always integer, the number of different rates that can be observed at any grid location depends on the number of births in the area. For filter areas with few infant births, the number of different rates will be few (Figure 5.3). In such cases there will be a large difference between the proportion of disease rates that can be less than the observed rate and the proportion that can be equal to or less than the observed rate. For example, if a filter area with 40 births had no deaths and 70% of simulated rates had no deaths, the percent of the simulated rates less than the observed rate is zero, whereas the percent of the simulated rates that were equal to or less than the observed rate is 70%. Thus, areas of a map can be defined as having infant mortality rates less than

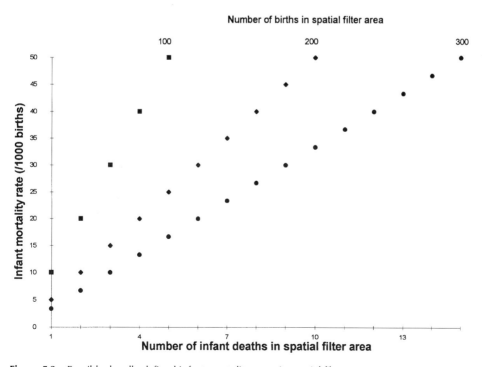

Figure 5.3 Feasible, locally defined infant mortality rates in spatial filter areas.

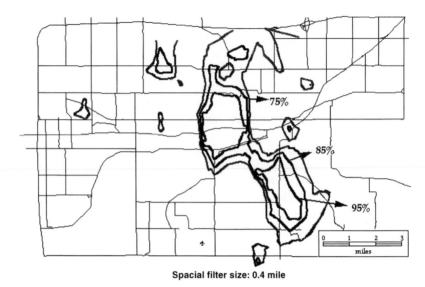

Figure 5.4 Significance levels for infant mortality rates less than or equal to the observed rate for a 0.4 mile filter area.

the observed rate (Figure 5.2a), or as having infant mortality rates that are equal to or less than the observed rate – Figure 5.4. Since the hypothesis of interest is the likelihood that what is observed in an area might have been produced by the process that generated the simulated rate, the proportion of simulated rates equal to or less than the observed rate is often of interest.

These maps contain, of course, multiple tests of significance, which means that at any given level of significance, some areas will be shown to be significant when they are really not – for example, false positives (Haybittle et al., 1995). This information – related to the statistical power of the test (Waller, 1996; Wartenberg and Greenberg, 1990; 1993) – must be weighed in interpreting the map of significance.

The ability of this method to detect patterns of significantly high disease rates was tested using it to analyse typical simulated disease patterns where significantly high rate areas should not be identified, given the null hypothesis of the Monte Carlo simulation that each birth had an identical probability of becoming a death. Other examples of such an approach to generating a reference distribution that reflects the known confounding factors in the population can be found elsewhere (Kingham et al., 1995, p. 812; Cuzick and Edwards, 1990, pp.78–79). In our studies of infant mortality rates in Des Moines, it was found that when the simulated patterns were examined to detect 'significantly' high rate areas – which would have been a spurious finding – the number of significantly high rate areas was more and their size much smaller than in the one significantly high rate in the observed pattern of infant deaths – see Figure 5.2 and Figure 5.4.

5.6 Computational issues in linking spatial analyses of disease and GIS

The steps followed in computing such maps are described in the flow chart (Figure 5.5). Three software programs are available on the CD-ROM (Rushton et al., 1996):

IMPROVING THE GEOGRAPHIC BASIS OF HEALTH SURVEILLANCE USING GIS 73

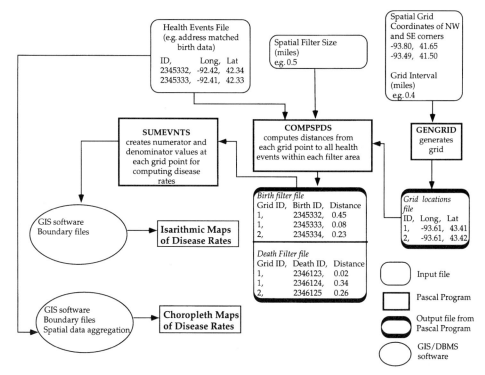

Figure 5.5 Steps in computing disease rate maps with variable spatial filters.

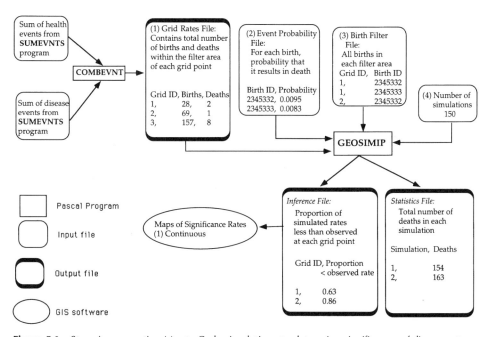

Figure 5.6 Steps in computing Monte Carlo simulations to determine significance of disease rates.

GENGRID generates a lattice of regular grid points and COMPSPDS computes distances from each grid point to all health events within each filter area. SUMEVNTS creates numerator and denominator values at each grid point for computing disease rates. Methods for applying spatial filters to point information are described elsewhere (Cliff and Haggett, 1988, pp. 36–37; Carrat and Valleron, 1992).The steps to compute such maps of significance are described in Figure 5.6. COMBEVNT creates the number of numerator and denominator events as computed previously with SUMEVNTS. The Monte Carlo simulation analysis is done by GEOSIMIP. This program creates the reference distribution and computes the proportion of the simulated rates at each grid point that are less than the observed rate. These files are input to a GIS for contouring. In these analyses, therefore, GIS software was used to address-match the birth and infant death records, visualise the grid and the characteristics of the filter area computations at each grid point, and compute and display the contours from the grid values. The computations described in Figure 5.5 and Figure 5.6 were completed outside of the GIS.

5.7 A decision support system for disease surveillance

Similar analyses to those described above can be performed for a wide variety of diseases and regions. The data required are widely available even though rarely used in this way. The software required for the analyses is available, as noted above, when used in conjunction with functions widely available in GIS software. The requirements are these:

1. Geo-coded disease data.
2. Geo-coded data that represent the population at risk. This may be an enumeration of persons at risk – as in this case where all births in the area constitute the geographical pattern of children at risk for infant death – or a suitably spatially stratified sample of people at risk as when the at-risk population may be represented by a census defined population group. For census defined groups, a suitable method of spatially disaggregating the population is required. Many GIS software have functions for this purpose.
3. Software for spatial filtering the disease and at-risk data – available in Rushton et al., 1996.
4. Software for contouring the filter area results – available in many but not all GIS software and also available in many statistical software.
5. Software for Monte Carlo simulation of disease rates and for comparing the observed rate at each grid point with the reference distribution of simulated rates at the same points – available in Rushton et al., 1996.

Since these requirements are frequently met it would be possible to bring these analyses steps together in an integrated software package that would perform the analyses for public health personnel. We call such a system a 'decision-support system (DSS) for geographically based health surveillance'. This suggested approach would be exploratory and interactive but would not aspire to the confirmatory goals envisioned by Openshaw (1994, p. 92) in which a Space–Time–Attribute Analysis Machine (STAM/1) is described for searching the three dimensions of space, time, and attributes.

Unlike Openshaw (1994, p. 93), we assume that the time and attribute space is known and, therefore, that the search can be restricted to geographic space. This considerably simplifies the far more general disease cluster search problem discussed by Openshaw. Such simplification, however, permits a richer search process that incorporates known risk factors in the Monte Carlo simulation experiments. It also permits computation of the variance in the test statistic due to the lack of independence between spatially contiguous measures. We view these two elements, the spatial autocorrelation between adjacent measures of disease rate – due to the sharing of observations in adjacent measures – and the incorporation of known risk factors in evaluating the spatial pattern of the disease, to be critical elements in the practice of public health surveillance. Key to the development of such knowledge-based spatial analysis systems is the ability to capture and encode the key decisions that analysts invoke as they proceed to analyse their data.

With access to GIS and the closely-coupled software on the CD-ROM, the means exist for every public health study to develop power studies for the unique conditions of their region. Traditionally, power studies were conducted to evaluate the relative power of alternative methods. Now, using GIS and supporting software, power studies can inform each study about the power of the test they have used in the context of their own conditions. With the increasing recognition that the power of any method is often conditional on the circumstances in which it is used (Cuzick and Edwards, 1990; McAuliffe and Afifi, 1984; Naus, 1966; Waller, 1996), the focus of attention in disease cluster investigations may shift from the power of the method to the power of the result in a given context. That is what is promised by these developments in local power tests of methods in their unique contexts. Recently, Waller (1996, p. 772) showed that the power of a test for clustering improved 'when the level of aggregation results in cells that are small enough to contain clusters without including areas with the background incidence rate'. This finding corroborates our intuition that the spatial aggregation approach of the choropleth map is a fundamentally flawed approach to assessing the statistical significance of disease rates whereas the spatially overlapping aggregations of Openshaw (1994), Hjalmars et al. (1996), and Rushton and Lolonis (1996) hold more promise for successfully identifying clusters, if they exist.

It is interesting to speculate on the consequences to the health of the public if such a DSS were developed and adopted in many areas. Presumably, the public would be interested in seeing and discussing the spatial patterns of disease in their area and, even though specialists in public health are well aware that local variations in disease rates are to be expected even when the process generating them assumes identical risk rates in the population, the public may not so readily accept such explanations and, in many cases, it may be quick to speculate that any number of factors may cause the disease pattern. The result may be that public health officials could become under siege to do something about high rates in local areas despite their spurious nature. Alternatively such widespread availability of a decision support system could become a vehicle for informing the public about typical patterns of disease rates and the tendency for local areas to have rates that fluctuate from year to year. They could become interested in public health programmes to improve the health of local populations and to become interested in monitoring the outcomes of such programmes and relating the geographic areas of improved outcomes to the areas of investment in health resources.

5.8 Conclusions

A transition from the traditional practice of disease cluster investigations that commence with putative disease clusters often nominated by the public and investigated at their behest, to regular geographic surveillance of disease patterns, will only occur if the public becomes far better informed and knowledgeable about the complex issues surrounding the analysis of disease patterns. The first step in this transition is to put exploratory spatial analysis methods into the hands of local public health professionals. The inefficiencies and other problems that are generated by traditional practice have been prominently explained by Rothman (1990). GIS permits the disease atlas to be complemented with an inference test for the detection of clusters. As Kulldorff and Nagarwalla (1995, p. 809) note, with such cluster detection methods, 'public health officials could better prioritize the regions with which to conduct thorough investigations, with minimization of the time taken to detect genuine abnormalities'.

Though the methods used in this chapter employ statistical tests for confirming disease clusters, they are, nonetheless, exploratory in purpose in their emphasis on the visual display of the geographic distribution of significance levels. Instead of identifying and viewing areas considered to be clusters at a given level of significance (Fotheringham and Zhan, 1996), this method provides the key information required in risk analysis. The method provides the information needed to implement local prevention and control programmes which is the primary purpose of public health surveillance systems.

Acknowledgments

This work is supported by grants from the US Department of Education and the Graduate College of the University of Iowa. Panos Lolonis wrote the software programs and Rajesh Krishnamurthy modified and incorporated them in the analysis system and prepared all illustrations for this paper.

References

BAILEY T. C. and GATRELL A. C. (1995) *Interactive Spatial Data Analysis*. Longman, Harlow.

BESAG J. and NEWELL J. (1991) The detection of clusters in rare diseases. *Journal of the Royal Statistical Society* Series A, **154**, 143–155.

BITHELL J. F. (1990) An application of density estimation to geographical epidemiology. *Statistics in Medicine*, **9**, 691–701.

BRILLINGER D. R. (1994) Examples of scientific problems and data analyses in demography, neurophysiology, and seismology. *Journal of Computational and Graphical Statistics*, **3**, 1–22.

CARRAT F. and VALLERON A-J. (1992) Epidemiologic mapping using the 'Kriging' method: application to an influenza-like illness epidemic in France. *American Journal of Epidemiology*, **135**, 1293–1300.

CHOYNOWSKI M. (1959) Maps based on probabilities. *Journal of the Royal Statistical Association*, **54**, 385–388.

CLIFF A.D. and HAGGETT P. (1988) *Atlas of Disease Distributions: analytic approaches to epidemiological data*. Blackwell, London.

CUZICK J. and EDWARDS R. (1990) Spatial clustering for inhomogeneous populations. *Journal of the Royal Statistical Society* Series B, **52**, 73–104.

DIGGLE P. J. (1991) A point process modelling approach to raised incidence of a rare phenomenon in the vicinity of a pre-specified point. *Journal of the Royal Statistical Society* Series A, **153**, 349–362.

DIGGLE P. J. and CHETWYND A. D. (1991) Second-order analysis of spatial clustering for inhomogeneous populations. *Biometrics*, **47**, 1155–1163.

DIGGLE P. J. and ROWLINGSON B. S. (1994) A conditional approach to point process modelling of elevated risk. *Journal of the Royal Statistical Society* Series A, **157**, Part 3, 433–440.

DOBSON A. J., KUULASMAA K., EBERLE E. and SHERER J. (1991) Confidence intervals for weighted sums of Poisson parameters. *Statistics in Medicine*, **10**, 457–462.

DUNCAN G. T. and PEARSON R. W. (1991) Enhancing access to microdata while protecting confidentiality: prospects for the future. *Statistical Science*, **6**, 219–239.

FOTHERINGHAM A. S. and ZHAN F. B. (1996) A comparison of three exploratory methods for cluster detection in spatial point patterns. *Geographical Analysis*, **28**, 200–218.

GATRELL A. C. and BAILEY T. C. (1996) Interactive spatial data analysis in medical geography. *Social Science and Medicine*, **42**, 843–855.

GATRELL A. C. and LÖYTÖNEN M. (1996) GIS and health research in Europe: a position paper. Paper prepared for the Helsinki workshop, January, 1996.

HAYBITTLE J., YUEN P. and MACHIN D. (1995) Multiple comparisons in disease mapping. *Statistics in Medicine*, **14**, 2503–2505.

HJALMARS U., KULLDORFF M., GUSTAFSSON G. and NAGARWALLA N. (1996) Childhood leukaemia in Sweden: Using GIS and a spatial scan statistic for cluster detection. *Statistics in Medicine*, **15**, 707–715.

INSTITUTE OF MEDICINE (1994) *Health Data in the Information Age: Use, Disclosure, and Privacy*. National Academy Press, Washington, D.C.

JACQUEZ G. M. (1996) Disease cluster statistics for imprecise space-time locations. *Statistics in Medicine*, **15**, 873–886.

JACQUEZ G. M. and WALLER L. A. (1996) The effect of uncertain locations on disease cluster statistics. *Proceedings of the Second International Symposium on Spatial Accuracy Assessment*, Fort Collins, Colorado, pp. 259–266.

KENNEDY-KALAFATIS S. (1995) Reliability-adjusted disease maps. *Social Science and Medicine*, **41**, 1273–1287.

KINGHAM S. P., GATRELL A. C. and ROWLINGSON B. (1995) Testing for clustering of health events within a geographical information system framework. *Environment and Planning A*, **27**, 809–821.

KULLDORFF M. and NAGARWALLA N. (1995) Spatial disease clusters: detection and inference. *Statistics in Medicine*, **14**, 799–810.

LANGFORD I. (1994) Using empirical Bayes estimates in the geographical analysis of disease risk. *Area*, **26**, 142–149.

LOVETT A. A., GATRELL A. C., BOUND J. P., HARVEY P. W. and WHELAN A. R. (1990) Congenital malformations in the Fylde region of Lancashire, England 1957–1973. *Social Science and Medicine*, **30**, 103–109.

MARSHALL R. J. (1991) A review of methods for the statistical analysis of spatial patterns of disease. *Journal of the Royal Statistical Society* Series A, **154**, 421–441.

MCAULIFFE T. L. and AFIFI A. A. (1984) Comparison of nearest neighbor and other approaches to the detection of space-time clustering. *Computational Statistics and Data Analysis*, **2**, 125–142.

MOULTON L. H., FOXMAN B., WOLFE R. A. and PORT F. K. (1994) Potential pitfalls in interpreting maps of stabilized rates. *Epidemiology*, **5**, 297–301.

NAUS J. I. (1966) A power comparison of two tests of non-random clustering. *Technometrics*, **8**, 493–517.

NEUTRA R., SWAN S. and MACK T. (1992) Clusters galore: insights about environmental clusters from probability theory. *The Science of the Total Environment*, **127**, 187–200.

ODEN N., JACQUEZ G. and GRIMSON R. (1996) Realistic power simulations compare point- and area-based disease cluster tests. *Statistics in Medicine*, **15**, 783–806.

OPENSHAW S. (1994) Two exploratory space-time-attribute pattern analysers relevant to GIS. In S. Fotheringham and P. Rogerson (eds) *Spatial Analysis and GIS*, Taylor & Francis, London, pp. 83–104.

OPENSHAW S., CHARLTON M. and CRAFT A. W. (1988a) Searching for leukaemia clusters using a geographical analysis machine. *Papers of the Regional Science Association*, **64**, 95–106.

OPENSHAW S., CHARLTON M., CRAFT A. W. and BIRCH J. M. (1988b) Investigation of leukaemia clusters by use of a geographical analysis machine. *Lancet*, **1**, 272–273.

OPENSHAW S., CHARLTON M., WYMER C. and CRAFT A. W. (1987) A Mark 1 geographical analysis machine for the automated analysis of point data sets. *International Journal of Geographical Information Systems*, **1**, 335–358.

REYNOLDS P., SMITH D. F., SATARIANO E., NELSON D. O., GOLDMAN L. R. and NEUTRA R. R. (1996) The four county study of childhood cancer: clusters in context. *Statistics in Medicine*, **15**, 683–697.

RIZZARDI M., MOHR M. S., MERRILL D. W. and SELVIN S. (1993) Interfacing U.S. Census map files with statistical graphics software: application and use in epidemiology. *Statistics in Medicine*, **12**, 1953–1964.

ROTHMAN K. J. (1990) A sobering start for the cluster busters' conference. *American Journal of Epidemiology*, **132**, S6–S13.

ROWLINGSON B. S. and DIGGLE P. J. (1993) SPLANCS: spatial point pattern analysis code in S-Plus. *Computers and Geosciences*, **19**, 627–655.

RUSHTON G., ARMSTRONG M. P., LYNCH C. and ROHRER J. (1996) *Improving Public Health Through Geographical Information Systems: An Instructional Guide to Major Concepts and Their Implementation*. The University of Iowa, Department of Geography, (CD-ROM), Iowa City, IA.

RUSHTON G. and LOLONIS P. (1996) Exploratory spatial analysis of birth defect rates in an urban population. *Statistics in Medicine*, **15**, 717–726.

SCHNEIDER D., GREENBERG M. R., DONALDSON M. H. and CHOI D. (1993) Cancer clusters: the importance of monitoring multiple geographic scales. *Social Science and Medicine*, **37**, 753–759.

SCHWEDER D. and SPJOTVOLL E. (1982) Plots of *P*-values to evaluate many tests simultaneously. *Biometrika*, **69**, 493–502.

SILVERMAN B. W. (1978) Choosing the window width when estimating a density. *Biometrika*, **65**, 1–11.

STONE R. A. (1988) Investigations of excess environmental risks around putative sources: statistical problems and a proposed test. *Statistics in Medicine*, **7**, 649–660.

STRASSBURG M. and WILLIAMS R. (1995) *EpiCMR User's Guide: An EpiInfo 6 application for importing data and generating reports, graphs, and maps from confidential morbidity reports*. Los Angeles County Department of Health Services, Los Angeles, CA; University of California, Community and Organization Research Institute, Santa Barbara, CA; April, 1995.

TURNBULL B. W., IWANO E. J., BURNETT W. S., HOWE H. L. and CLARK L. C. (1990) Monitoring for clusters of disease: application to Leukemia incidence in upstate New York. *American Journal of Epidemiology*, **132**, S136–S143.

TWIGG L. (1990) Health-based geographical information systems: their existing potential examined in the light of existing data sources. *Social Science and Medicine*, **30**(1), 143–155.

US BUREAU OF THE CENSUS (1992) *TIGER/Line Files, 1992. Technical Documentation*. The Bureau, Washington D.C.

WALLER L. A. (1995) Statistical power and design of focused clustering studies. *Statistics in Medicine*, **15**, 765–782.

WALLER L. A. and TURNBULL B. W. (1993) The effect of scale on tests for disease clustering. *Statistics in Medicine*, **12**, 1869–1884.
WARTENBERG D. and GREENBERG M. (1990) Detecting disease clusters: the importance of statistical power. *American Journal of Epidemiology*, **132**, S156–S166.
WARTENBERG D. and GREENBERG M. (1993) Solving the cluster puzzle: clues to follow and pitfalls to avoid. *Statistics in Medicine*, **12**, 1763–1770.
WEINSTOCK M. A. (1981) A generalized scan statistic test for the detection of clusters. *International Journal of Epidemiology*, **10**, 289–293.

CHAPTER SIX

Modelling Spatial Variations in Air Quality using GIS

SUSAN COLLINS

6.1 Introduction

There is widespread concern about increases in the number of vehicles on the roads and how this may relate to public health. Increases in numbers of vehicles and emissions from road transport have been well documented (Commission of the European Communities, 1992; Quality of Urban Atmospheric Review Group, 1993). Recent articles published in the medical journals have expressed concern over increases in respiratory diseases and the association with air pollution (Dockery *et al.*, 1993; Pope *et al.*, 1995; Schwartz, 1993). A number of studies have found a relationship between traffic related indicators, for example, proximity to major roads, living near roads with high volumes of traffic, and health effects (Wieland *et al.*, 1994; Murakami *et al.*, 1990; Edwards *et al.*, 1994; Wjst *et al.*, 1993; Nitta *et al.*, 1993). More recently the Committee on the Medical Effects of Air Pollution have produced a report to advise on the possible links between outdoor air pollution and asthma (Department of Health, 1995) and expert panels have been commissioned by the Department of the Environment (1995a, b, c) to recommend air quality standards for a number of pollutants, including benzene, carbon monoxide and particles.

This research activity highlights the need to control and reduce air pollution in major cities. It is therefore important to identify areas where pollution levels exceed guidelines and standards. Thus policy and planning strategies can then be targeted at those areas where pollution levels are high, identifying areas where the public may be at risk to health effects and where regulatory measures would be beneficial to the public.

Against this background it is essential that reliable predictions of air quality can be derived. Models of the spatial variations in air quality are therefore needed if accurate and reliable predictions are to be established. The models can then be used to produce detailed and accurate maps of air pollution.

Models can be used to forecast future levels in air quality to inform the design and implementation of control strategies. Maps of air quality can be used to help design monitoring networks – selecting locations for new monitoring sites – and they can also be used to identify areas that do not comply with EC directives. The maps will help to identify the major emission sources and monitor the effects of policy once it is implemented.

Maps have an important impact from a health perspective. Epidemiologists looking at the geographical distribution of health are looking at large populations. The ideal way to estimate an individual's exposure would be by personal monitoring. In large population surveys, however, where you are perhaps looking at a few thousand individuals – this would be, not only extremely complex to manage and monitor, but also extremely expensive and time-consuming. Maps therefore offer an effective solution to estimate individual exposure and help establish associations between pollution and health.

6.2 Air quality mapping

One way of establishing pollution concentration at unsampled locations is dispersion modelling. These models are often developed as a function of the Gaussian dispersion curve. The models use information related to emission sources of pollution and characteristics of the dispersion environment. For example, in the case of line dispersion models this might include: composition of traffic, traffic volume, emission rates, traffic idling time, meteorology, mixing height and information about the surface adjacent to the road, such as housing density. The coordinates of a road link (road segment of uniform conditions) and the coordinates of receptors (locations around the road link for which concentrations are calculated) are entered into the model. For both point and line dispersion models, a limited number of multiple sources can often be entered into the model.

The models can be used either to calculate pollution concentration at specific locations (for example the place of residence of individuals) or to produce an array of points that can be entered into a mapping package and a continuous surface generated by interpolation. In the case of line dispersion models, creating a continuous surface is not that simple. Variations in the characteristics of a road change quite rapidly in urban areas, for example, due to: changes in the housing density at the side of a road; variations in traffic volume and speed; and changes in the direction of the road. Consequently in area-wide studies a great many road segments need to be modelled. Where a whole city has to be modelled, the demands on data are therefore very high. Furthermore the models can only accurately estimate pollution levels close to the source, 35 m in the case of CAR (Eerens *et al.*, 1993) and 200 m in the case of CALINE (Benson, 1992). In the CAR model, a city-wide background concentration is calculated, however, most line dispersion models do not estimate background levels of pollution.

6.2.1 Air quality monitoring

An alternative to dispersion modelling in area-wide studies is air quality monitoring. At any one location there may be pollution from a variety of different sources effecting that point. It is a difficult task to try to identify and model all the sources. The advantage of monitoring is that it measures the contribution of air pollution from all sources that have an effect on a location.

Air quality in the UK is measured at 47 automatic stations (Bower and Vallance-Plews, 1995). The pollutants that are measured include O_3, NO_2, SO_2, CO, PM_{10} and Speciated HC, although not all pollutants are measured at all sites. Automatic stations enable short-term measurements of exposure to be undertaken and therefore can record specific pollution events. Unfortunately they are very expensive to purchase and operate, and consequently there are very few sites – although more sites are planned for the future.

At least for one of the pollutants associated with vehicle emissions, NO_2, the diffusion tube offers an alternative measuring device. A diffusion tube is a passive sampler which measures the time weighted average concentration of NO_2 in ambient air. The most commonly used sampler is the Palmes tube (Palmes et al., 1976) which is a passive diffusion device consisting of an acrylic tube open at one end with coated stainless steel screens at the closed end. The NO_2 diffuses through the air in the tube and is trapped as nitrite ions in the triethanolamine which coats the screens.

The tubes are relatively inexpensive and hence facilitate a higher density of sample locations, providing more detailed information. The tubes are usually exposed for periods of 1 to 4 weeks, and measure the mean NO_2 concentration over that period (Her Majesty's Inspectorate of Pollution, 1993). The tubes are not really sensitive for periods of less than one week and therefore can only reliably be used to look at the chronic (long-term) health effects of NO_2.

The diffusion tube is a simple and cost-effective device, consequently NO_2 is the pollutant most commonly measured as a proxy for the complex of traffic related pollutants. Through government funded national monitoring programmes there are now 1200 sample sites for NO_2 as part of the UK monitoring networks (Bower and Vallance-Plews, 1995). In addition, many local authorities are using diffusion tubes for local monitoring surveys.

6.2.2 Kriging

Maps of continuous variables are usually generated by interpolating from a sample set of points. Many interpolation techniques exist. The geostatistical technique kriging, however, has been found to be the most effective in the majority of environmental applications. Kriging is an optimal interpolator whose estimates are unbiased and have known minimum variance (Oliver and Webster, 1990). The technique is based upon the theory of regionalised variables and utilises the spatial structure of the data. It involves the construction of a variogram and the fitting of an appropriate model. The kriging interpolation estimates by local weighted averaging, where the weights have been determined by the variogram and the configuration of the data (Oliver and Webster, 1990). More information about kriging and discussions on the role of kriging in the environmental sciences can be found in Haining (1990) and Bailey and Gatrell (1995).

Recently, kriging has been applied in studies of air quality, where area-wide levels of air pollution have been estimated from values recorded at monitoring sites. For example, Atkins and Lee (1995) estimated rural levels of NO_2 for the UK using kriging and Liu et al. (1995) examined the use of kriging and co-kriging to estimate levels of ozone concentration in Toronto, USA.

6.2.3 Variations in NO_2

In urban areas variations in levels of NO_2 have been shown to occur over very small distances. For example, research undertaken by Laxon and Noordally (1987) to investigate the distribution of NO_2 in street canyons, suggest that NO_2 concentrations decline rapidly from the centre of the road with concentrations close to local background levels at a distance of 30 m, and with nearly two-fold variations in NO_2 with distances less than 100 m. High degrees of local variation have also been found by Hewitt (1991), who examined spatial variations of NO_2 in Lancaster, UK, based on diffusion tube measurements. As these studies demonstrate motor vehicle emissions dominate NO_2 concentrations in cities and near to roads, with vehicle emissions strongly influencing spatial variations in NO_2 at ground level.

Although kriging is a sophisticated interpolation technique, line pollution sources, as opposed to point pollution sources, are a lot more difficult to represent by interpolation methods. The nature of line-source pollution is that it will peak strongly where the main roads are, and unless the monitoring sites have been placed next to the sources the interpolation will smooth the variation that occurs between the points.

It is evident, therefore, that NO_2 needs to be modelled at a resolution that reflects these spatial variations. This need is supported by the fact that policy and planning strategies are implemented at a local level and that the majority of health data is supplied at a high resolution, that is postcode level. If an individual has been located by his/her postcode, then it seems sensible to maintain that level of detail when trying to estimate the potential 'risk'.

6.3 The role of GIS

GIS are sophisticated software which allow the fast automation of historically complex routines. On a simple level GIS can be used automatically and efficiently to calculate health indicators, for instance, for traffic pollution: distance to nearest road, traffic volume on nearest road or an impression of decay using some weighted composite value of the two, for example, traffic volume over distance. Although these types of exposure indicators are a useful estimate, they can be misleading. For example, in the case of traffic volume on the nearest main road, the nearest main road is not necessarily the primary source of pollution, there may be another road just behind the first that could be carrying two or three times as much traffic.

Another important role of GIS is the ability to bring together data from a variety of sources, for example, health, socioeconomic data and environmental data, within a common framework. Once a map has been created in the GIS it can be integrated with other information, such as census data. Gatrell and Löytönen (Chapter 1) identify the integration of environmental modelling and health databases, within a GIS, as a primary area of research.

Many GIS now have the capacity to store and analyse data in grid format. The grid has evolved from remote sensing and a lot of the tools that can be used to analyse data were originally used to correct, enhance and interpret remotely sensed images. What the grid offers the user is a powerful tool for modelling spatial variations and the capacity to analyse data at a fine resolution. This is extremely useful for modelling environmental phenomena, such as air quality, which do not vary according to administrative and political boundaries, or some other aggregation.

GIS can thus be effectively used to produce more realistic estimates of exposure by generating models of air quality. As digital geographical data become more accessible and affordable, many local authorities, government departments and environmental agencies now have access to these data. The large volumes of spatially referenced data stored in the GIS, at a fine resolution, can usefully be used to help produce more realistic models of the spatial distribution of air quality and hence, more reliable estimates. The same arguments apply for modelling many other environmental phenomena.

Two methods to model air quality have been developed using GIS, the first is a hybrid approach that combines dispersion modelling with spatial interpolation and the second is a regression approach that combines GIS tools and statistical techniques. These methods are described in detail below and are compared with the geostatistical technique kriging.

6.4 The SAVIAH study

The methods were developed as part of the Small Area Variations in Air Quality and Health (SAVIAH) project. The SAVIAH project was an EU-funded methodological study aimed to apply, test and evaluate new and emerging methodologies in the fields of epidemiology, geography, air pollution modelling and small area health statistics, and to bring the data together in a consistent geographic framework (Elliott *et al.*, 1995). The methods were developed using data collected for Huddersfield, UK, a mixed urban and rural area. Huddersfield was one of four study areas in the project. The study brought together, in a common geographical framework within a GIS, three major data sets. These are discussed in turn.

6.4.1 Air pollution data

Nitrogen dioxide was measured during four survey periods of two week duration in June 1993, October 1993, March 1994 and June 1994 (Smallbone, 1997). Nitrogen dioxide was measured as a proxy for the complex of traffic related pollutants and was chosen for its ease of measurement. The measurements were undertaken using Palmes diffusion tubes, placed at 80 'permanent' and 40 'variable' sites for each survey. Permanent sites were locations which were the same for each survey, whereas variable sites were relocated for each survey to examine specific patterns and sources of variation. In addition, 8 consecutive sites were exposed for the full duration of the study period on a monthly cycle, apart from when the individual surveys were undertaken, when they were exposed for the same two week period. The permanent sites were used to develop the models and the consecutive sites were used to validate the models. Two tubes were exposed at all sites to provide a measure of the at-site variation and also to provide insurance against loss or damage of tubes. In the study approximately 10% of the measurements were lost through vandalism or damage. To compensate for this loss of data, mean concentrations were established for the permanent sites using multilevel modelling techniques, with terms for measurement error and site and survey effects (Lebret *et al.*, 1995).

The locations of the monitoring sites were chosen to reflect the population and the sources of pollution. Additional sites were chosen in background locations, away from the sources of pollution to act as a control and help establish the background level of NO_2. Further consideration was given to the spatial distribution of the sites: where possible sites were evenly spread across the study area, avoiding excessive distances between neighbouring points; with sites denser in the densely populated areas; and at most roadside locations one site was placed at the kerbside and another 50–100 m back from the kerb.

At each site field information was also recorded, this included height of the sample above the ground and the vertical angle to the horizon.

6.4.2 Health data

All children aged between 7 and 11 attending a school in the Huddersfield study area were interviewed by questionnaire. There was a 91% response rate which resulted in information for approximately 4500 children.

The questionnaire included questions about the home (for example, heating, damp, pets), questions about the parents (for example, education, smoking habits, respiratory disorders), address and postcode, length of time at that address, previous address if valid, school of the child and 15 questions related to the health of the child – with particular reference to respiratory disorders.

6.4.3 Geographical data

Geographical data were collected for the road network (with attributes for type of road and traffic volume) and land cover (including 7 urban classes) and a digital terrain model was created.

6.5 Hybrid approach

The hybrid approach attempts to link dispersion modelling to the GIS. The advantages of linking line dispersion models to the GIS are:

- High resolution modelling.
- Many road segments can be modelled automatically – this is especially useful for large study areas.
- The user doesn't need to transfer data from one system to another – running the dispersion model outside the GIS and then transferring the results back into the GIS for mapping.
- The user doesn't need to learn two different software packages (probably running on two different operating systems).

The approach uses a combination of dispersion modelling and geostatistical techniques available in the GIS. The dispersion modelling aims to describe the near-source variation related to dispersion processes associated with the line sources. Geostatistical techniques are employed to model the background variation, related to other sources and controls on the dispersion of the pollutant, and the long-distance transport of pollutants.

To predict the near-source variation the line dispersion model CALINE3 has been adapted to work within the GIS. The method is described in detail elsewhere (Collins *et al.*, 1995a; Collins, 1996).

Briefly, CALINE3 was run many times for one road link with 6 receptors on each side of the road, perpendicular to the road and at varying distances away from the road, up to 200 m. The model was run with different combinations of wind direction, wind speed, atmospheric stability, road type and surface roughness. Each run of the model described one possible scenario of conditions. A constant traffic volume and emission factor were entered into the model. For each scenario the pollution concentration at the 12 receptors was recorded. The results were stored in a look-up table in the GIS. Each record in the table relates to one scenario. The columns in the table include: a unique 5 digit code that identifies a particular scenario, 6 columns for the receptors on the downwind side of the road and 6 columns for the receptors on the upwind side of the road.

A program was written to calculate automatically the concentration level of pollution in the near-source area (200 m from the main roads) on a 10 m grid in the GIS. The program used the geographical information stored in the GIS (main road network, type of road, traffic volume and land cover) and a file containing the meteorological data for the survey period (wind direction, wind speed and atmospheric stability). For all cells in the near-source area the following information was found: orientation of the road (with respect to wind direction); traffic volume on the road; distance to the road and surface roughness (a function of land cover). This information, along with the meteorological data, was then used to establish which receptor corresponded to the cell, if the cell was upwind or downwind and a 5 digit numerical code equivalent to one of the unique codes in the look-up table. The concentration values could then be extracted from the look-up table and the concentration proportionally adjusted for traffic volume.

Ordinary kriging was applied to those permanent monitoring sites that fell in the background areas, that is beyond 200 m, to create a continuous surface of background pollution. The near-source pollution was then added to the background pollution to create a final map. For the monitoring sites not used in the model, that is those that fell in the near-source areas, the model was found to explain 52% of the variation.

Although this is an improvement on kriging and allows the near-source variation to be more accurately defined, it does have some disadvantages. The method is still quite data intensive and is also computationally expensive. It was found, however, that for measuring the long-term (annual) pollution concentrations, the additional data relating to meteorology did not significantly improve the accuracy of the model. Therefore, over long periods of time, the effects due to meteorology become negligible. When the model is run with an annual average wind rose for wind direction and annual mean values for wind speed and atmospheric stability the method becomes computationally more efficient and less data intensive.

Kriging the background monitoring sites, however, is not always a viable solution. Producing the variogram may indicate that there is no spatial dependence in the data, this could be due to too few monitoring sites in the background area or a poor distribution of sites.

6.6 Regression approach

The regression approach uses a combination of GIS and statistical techniques and was developed to overcome the problems identified with the hybrid approach. It is based upon the idea that pollution at a particular location is a product of the pollution from all sources in the surrounding area and that nearby sources are likely to contribute more than distant sources.

The method is more easily transferred to other areas and is relatively simple to apply. Consequently, the regression approach was adopted in other study areas in the SAVIAH project. The methodology, implementation and evaluation of the approach have been reported in detail in Briggs et al. (1996), Collins et al. (1995b) and Collins (1996) and are summarised briefly below.

Factors that reflect the variations in air pollution at street level were identified. These fell into two categories: sources of pollution and factors that reflect the dispersion patterns and rates of pollution. The predominant source of air pollution is emissions from motor vehicles, but there are also contributions from industrial and domestic activities, and building density and topography determine the patterns and rates of pollution. In this model traffic volume, land cover and topography are used as proxies for these factors.

In the first instance an indicator for traffic volume was identified. The total daytime hourly traffic volume on roads within successive 20 m bands up to 300 m radius (that is 0–20 m, 20–40 m, 40–60 m) around the permanent monitoring sites were found. In the statistical package SPSS, the adjusted mean annual NO_2 concentrations for the permanent monitoring sites were entered into a multiple regression analysis against different combinations of band width. By examination of the R^2 value the best-fit combination was identified – this proved to be the 0–40 m and the 40–300 m bands. Weights for the bands were defined by the slope coefficients which gave a ratio of 15 to 1. This gave the following compound indicator:

$$\text{TRAVOL} = (15 * \text{traffic volume 0–40 m band})$$
$$+ (1 * \text{traffic volume 40–300 m band})$$

An indicator reflecting the built environment was then established. The total area of different urban land cover categories was again found for successive 20 m bands up to 300 m radius. Different combinations of bands were entered into a multiple regression analysis against the residuals from the compound traffic volume indicator. Again, the best combination was identified by examination of the R^2 value and the weights defined by the slope coefficients to give the following compound indicator:

$$\text{LANDCOV} = (1.8 * \text{high density residential 0–300 m band})$$
$$+ (1 * \text{industry 0–300 m band})$$

The compound indicators were then entered into a stepwise multiple regression analysis, with altitude (transformed) and sample height against the adjusted mean NO_2 values (the dependent variable) for the 80 permanent sites. The final regression equation was found to be:

$$\text{Mean } NO_2 = 11.83 + (0.00398 * \text{TRAVOL}) + (0.268 * \text{LANDCOV})$$
$$- (0.0355 * \text{SIN(altitude)}^{-1}) + (6.777 * \text{sample height})$$

The equation was then applied in the GIS, on a cell by cell basis, to estimate NO_2 for all locations across the study area at a 10 m resolution.

All variables were found to be significant at the 0.005 level and the equation was found to explain 62% of the variation. Analysis of the residuals did not reveal any further spatial dependence in the data. The approach was applied in three different study areas and was found to be an effective and reliable technique in all three (Briggs et al., 1996).

6.7 Comparison of the different approaches

The different approaches were compared by predicting mean NO_2 for the 8 consecutive monitoring sites. A graph of annual mean NO_2 against predicted NO_2 for kriging, the hybrid approach and the regression approach can be seen in Figure 6.1 and the R^2 values, constant and slope coefficient can be seen in Table 6.1. In the hybrid approach, one of the monitoring sites was identified as an outlier, in the second hybrid entry the outlier has been removed.

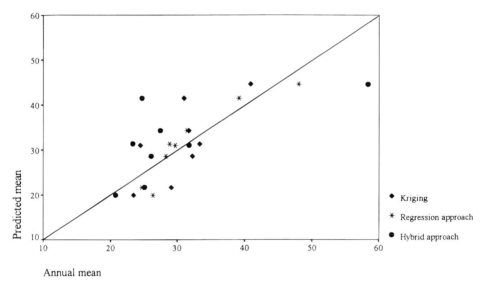

Figure 6.1 Observed and predicted NO_2 (ugm^{-3}): kriging, hybrid and regression approaches, 8 consecutive sites.

As can be seen from the graph and the table, the regression approach proves to be the best predictor, with the highest R^2 value of 81.7%. Figures 6.2, 6.3 and 6.4 show the final pollution maps for the kriging, hybrid and regression approaches

Table 6.1 Adjusted R^2, constant and slope coefficients for the kriging, hybrid and regression approaches

Method	Adjusted R^2	Constant	Slope coefficient	Standard error of the estimate
Ordinary kriging	43.9%	−3.44	1.14	6.45
Hybrid	62.8%	13.21	0.59	5.25
Hybrid*	60.6%	−1.07	1.11	4.63
Regression	81.7%	−0.60	1.02	3.68

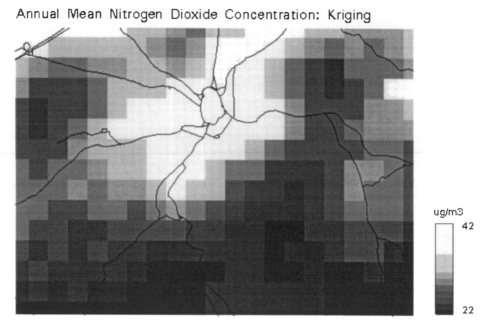

Figure 6.2 Annual NO_2 concentration produced by kriging.

respectively. When a pollution map is combined with other information, such as population or health data, results are very dependent upon the choice of model, and therefore must be interpreted with caution. To demonstrate the extent of this problem, a threshold pollution value for the study area was calculated by finding

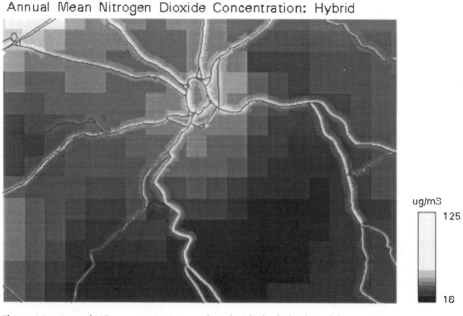

Figure 6.3 Annual NO_2 concentration produced with the hybrid model.

Annual Mean Nitrogen Dioxide Concentration: Regression

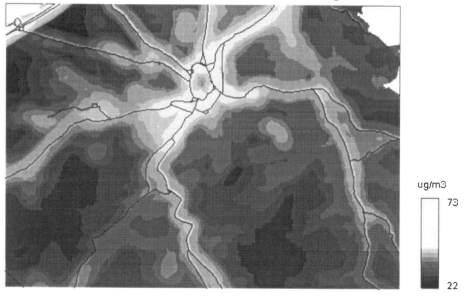

Figure 6.4 Annual NO$_2$ concentration produced by the regression model.

Residential Areas

Figure 6.5 Residential areas.

Table 6.2 Minimum and maximum pollution values for residental areas and the percentage of residential area above the average monitored pollution level

Method	Minimum residential value	Maximum residential value	Residental area > 33.06
Ordinary kriging	21.64	41.67	35%
Hybrid	18.15	82.45	9%
Regression	23.06	58.16	17%

the average of the adjusted annual mean NO_2 concentrations for the 80 permanent monitoring sites. This was found to be 33.06. The residential areas, taken from the land cover map (shown in Figure 6.5), were then overlaid with the three pollution maps. The percentage of residential area that fell above the threshold value was calculated for all three models. The results, along with the minimum and maximum pollution values in the residential areas, are shown in Table 6.2.

As can be seen from the table, the percentage of residential area above the threshold value changes quite dramatically for the different models. At one extreme, the kriging model suggests that over a third of residential areas have pollution levels above the monitored average, while at the other extreme the hybrid model indicates that less than one tenth of residential areas fall in this category. In the case of kriging, this high percentage, and the small range of values between the minimum and maximum scores compared with the other models, demonstrates how the kriging technique has smoothed the pollution surface. It is evident, therefore, that reliable and accurate estimates of air quality are essential if the true impact of air pollution on public health is to be assessed.

6.8 Discussion and conclusions

Very few sample sites can be established where equipment to monitor air pollution is expensive to purchase and operate. Hence, sites are often located where levels of pollution are expected to be high. This results in a poor representation of the pollution surface.

Low-cost monitoring such as diffusion tubes, which are simple and cheap to operate, provide a higher density of sample sites, affording a more reliable representation of the pollution surface. Establishing estimates at unsampled locations by monitoring and then interpolating can therefore be undertaken quite cheaply and with little data requirement. A manageable number of monitoring sites can therefore be effectively used to undertake small localised surveys, for example, road transects, a street junction or residential locations for a sample population.

In the case of large population studies where estimates are required for many unsampled locations, interpolating from monitoring sites is not such a viable option. Spatial variations in air pollution from vehicle emissions occur over very small distances. It would require an unrealistic density of monitoring sites to pick up all the local variation that occurs. Consequently, when a limited number of sites are used the local variation is smoothed. This can result in misrepresentation of the data, which may cause an over-estimation in population exposure and an under-estimation of the severity of exposure at extreme locations.

A map produced from monitored data will only reflect the measured values and the time period for which the measurements were taken. With the introduction of modelling, it is possible to predict variations in pollution, including extreme values (that is peaks and troughs), and introduce daily and seasonal variations into the model, without additional monitoring. In this manner, past and future levels of pollution can be estimated.

This is extremely useful for policy makers and planners, who are interested in the pattern and pathways of pollution at ground level. For example, models can be used to assess the impact of new schemes, such as building a new road or giving restricted access to an existing road, before implementation, and monitor the effect on levels of pollution after implementation.

The hybrid and regression approaches have shown that information stored in the GIS can be used to help produce better maps and therefore more reliable estimates of exposure. The two approaches also demonstrate that this can be undertaken with only a small and manageable number of monitoring sites.

GIS have the tools to develop good, high resolution models of air quality. Although the regression model proved to be the best approach, data still had to be transferred from the GIS, the statistical analysis applied in another package, and the results imported back into the GIS. The inability of GIS to perform statistical analysis, such as regression and multiple-regression, on items stored in the database is a severe limitation of all GIS packages. The introduction of these, and other statistical techniques, would greatly enhance the modelling capacity of GIS.

The link between vehicle emissions and health is extremely complex. Attempting to quantify and measure exposure – not least define it – is a very difficult and often controversial task. People are very mobile and individuals are exposed to different levels and combinations of pollutants. If we wish to understand the association between air quality and health, then we need to understand and model, not only the spatial variation of pollution, but also the activities and mobility of the population.

Acknowledgments

Some of the research presented in this chapter was undertaken as part of the EU-funded SAVIAH project. The author would like to thank all those involved in the SAVIAH project, particularly Professor David Briggs (Nene Centre for Research, Nene College, Northampton, UK) and Kirsty Smallbone (Brighton University, UK).

References

ATKINS D. H .F. and LEE D. S. (1995) Spatial and temporal variations of rural nitrogen dioxide concentration across the United Kingdom. *Atmospheric Environment*, **29**, 223–239.

BAILEY T. C. and GATRELL A. C. (1995) *Interactive Spatial Data Analysis*. Longman, Harlow.

BENSON P. E. (1992) A review of the development and application of the CALINE3 and CALINE4 models. *Atmospheric Environment*, **26B**, 379–390.

BOWER J. S. and VALLANCE-PLEWS J. (1995) The UK national air monitoring networks. *Paper presented at a WHO seminar*, 21–23 November 1995.

BRIGGS D. J., COLLINS S., ELLIOTT P., FISCHER P., KINGHAM S., LEBRET E., PRYL K., VAN REEUWIJK H., SMALLBONE K. and VAN DER VEEN A. (1996) Air pollution mapping in the SAVIAH study. Submitted to *International Journal of GIS*.

COLLINS S. (1996) Modelling urban air pollution using GIS. In Craglia M. and Couclelis H. (eds) *Geographic Information Research: Bridging the Atlantic*, Taylor and Francis, London.

COLLINS S., SMALLBONE K. and BRIGGS D. (1995a) A GIS approach to modelling small area variations in air pollution within a complex urban environment. Pages 245–253 in Fisher P. (ed.) *Innovations in GIS 2*, Taylor and Francis, London.

COLLINS S., SMALLBONE K. and BRIGGS D. (1995b) A regression model for estimating small area variations in air pollution. *Epidemiology*, **6**(4), S60.

COMMISSION OF THE EUROPEAN COMMUNITIES (1992) *Towards Sustainability: a Program of Action on the European Environment*. Commission of the European Communities, Brussels.

DEPARTMENT OF HEALTH (1995) *Committee on the Medical Effects of Air Pollutants. Asthma and Outdoor Air Pollution*. HMSO, London.

DEPARTMENT OF THE ENVIRONMENT (1995a) *Expert Panel on Air Quality Standards. Benzene*. HMSO, London.

DEPARTMENT OF THE ENVIRONMENT (1995b) *Expert Panel on Air Quality Standards. Carbon Monoxide*. HMSO, London.

DEPARTMENT OF THE ENVIRONMENT (1995c) *Expert Panel on Air Quality Standards. Particles*. HMSO, London.

DOCKERY D. W., POPE C. A. III, XU X., SPENGLER J. D., WARE J. H., FAY M. E., FERRIS B. G. JR and SPEIZER F. E. (1993) An association between air pollution and mortality in six US cities. *Journal of Medicine*, **329**, 1753–1759.

EDWARDS J., WALTERS S. and GRIFFITHS R. K. (1994) Hospital admissions for asthma in preschool children: relationship to major roads in Birmingham, United Kingdom. *Archives of Environmental Health*, **49**, 223–227.

EERENS H., SLIGGERS C. and VAN DER HOUT K. (1993) The CAR model: the Dutch method to determine city street air quality. *Atmospheric Environment*, **27B**, 389–399.

ELLIOTT P., BRIGGS D., LEBRET E., GORYNSKI P. and KRIZ, B. (1995) Small Area Variations in Air Quality and Health: the SAVIAH study. *Epidemiology*, **6**, S31 (Abstract).

HAINING R. P. (1990) *Spatial Data Analysis in the Social and Environmental Sciences*. Cambridge University Press, Cambridge.

HER MAJESTY'S INSPECTORATE OF POLLUTION (1993) *An Assessment of the Effects of Industrial Releases of Nitrogen Oxides in the East Thames Corridor*. HMSO, London.

HEWITT C. N. (1991) Spatial variation in nitrogen dioxide concentrations in an urban area. *Atmospheric Environment*, **25B**, 429–434.

LAXEN D. P. H. and NOORDALLY E. (1987) Nitrogen dioxide distribution in street canyons. *Atmospheric Environment*, **21**, 1899–1903.

LEBRET E., BRIGGS D., COLLINS S., VAN REEUWIJK H. and FISHER P. H. (1995) Small area variation in exposure to NO_2. *Epidemiology*, **6**, S31.

LUI L., SALLY J., ROSSINI, A. and KOUTRAKIS P. (1995) Development of cokriging models to predict 1- and 12-hour ozone concentrations in Toronto. *Epidemiology*, **6**, S69.

MURAKAMI M., ONO M. and TAMURA. K. (1990) Health problems of residents along heavy-traffic roads. *Journal of Human Ergol*, **19**, 101–106.

NITTA H., SATO T., NAKI S., MAEDA K., AOKI S. and ONO M. (1993) Respiratory health associated with exposure to automobile exhaust: I. Results of cross-sectional study in 1979, 1982 and 1983. *Archives of Environmental Health*, **48**, 53–58.

OLIVER M. A. and WEBSTER R. (1990) Kriging: a method of interpolation for geographical information systems. *International Journal of GIS*, **4**, 313–332.

PALMES E. D., GUNNISON A. F., DIMATTIO J. and TOMCZYK C. (1976) Personal sampler for nitrogen dioxide. *American Industrial Hygiene Association Journal*, **37**, 570–577.

POPE C. A. III, THUN M. J., NAMBOODIRI M. M., DOCKERY D. W., EVANS J. S., SPEIZER F. E. and HEATH C. W. JR (1995) Particulate air pollution as a predictor of mortality in a prospective study of US adults. *American Journal of Respiratory and Critical Care Medicine*, **151**, 669–674.

QUALITY OF URBAN ATMOSPHERIC REVIEW GROUP (1993) *Urban Air Quality in the United Kingdom*. First Report of the Quality of Urban Atmosphere Review Group, DoE, Bradford.

SCHWARTZ J. (1993) Particulate air pollution and chronic respiratory health. *Environmental Research*, **62**, 7–13.

SMALLBONE K. (1997) Unpublished PhD thesis.

WIELAND S. K., MUNDT K. A., RUCKMANN A. and KEIL U. (1994) Reported wheezing and allergic rhinitis in children and traffic density on streets of residents. *AIR*, **4**, 79–84.

WJST M., REITMEIR P., DOLD S., WULFF A., NICOLAI T., VON LEOFFELHOLZ-COLBURG E. and VON MUTIUS E. (1993) Road traffic and adverse effects on respiratory health in children. *British Medical Journal*, **307**, 596–600.

CHAPTER SEVEN

GIS, Time Geography and Health

MARKKU LÖYTÖNEN

7.1 Exposure and time–space lag

At an individual level, we are used to observing time-dependent changes such as ageing in ourselves and in people close to us. Diseases occurring in childhood are predominantly infections and congenital malformations. With increasing age, degenerative processes and malignant neoplasms become more dominating causes of ill-health in people living in industrial countries. What we tend to forget is that changes in our living environment during the average life time typical for the industrial world – over 70 years – are often equally substantial. Although some of the changes in the living environment can draw much attention (such as the nuclear accident at Chernobyl) most changes develop gradually and remain largely unnoticed. When an individual becomes ill, in most cases the modern health care system focuses on the individual, with little interest in establishing epidemiologically important aetiological factors, some of which may be distant both in terms of time and location. Yet understanding the individual's time–space history can provide important information not only for the epidemiologist, but also for the clinician.

Irrespective of the aims of the research or of the disease which forms its object, the material will always consist of information on those suffering from the disease. Information on the disease contracted by the population concerned, for example its nature, incidence, or stage, constitutes the attribute data and the information on the places where these people live, for example map coordinates, postal area codes, or local government districts, the georeferenced data. Regardless of whether the research is genuinely based on individual data or whether it relies on aggregate data for given areal units, our notions of prevalence and incidence and the spatial and temporal variations in these will always be based on information on individual cases recorded at the time of diagnosis. The possibility of using georeferenced register data at the individual level and the possibility of combining these with data from other registers via the social security numbers of the people concerned opens up opportunities for a wide range of accurate spatial analyses. Spatially, temporally and medically reliable and accurate though these sets of data may be, they nevertheless entail certain problems, both geographical and medical, which complicate the research process.

From a geographical point of view, the place of residence recorded for the patient is the place where he or she is living at the time when the disease is diagnosed, even though in practice the person may only just have moved to that area or may be registered as resident in a different administrative district. Since we know that a considerable proportion of the population of the industrialised countries move from one district to another within the country, or even from one country to another, in the course of a year, it is obvious that this can introduce an element of inaccuracy which is extremely difficult to evaluate but which can have a notable effect on the results of geographical research into matters connected with the health of the population (Bentham, 1988).

Considerably more problematical, however, is the question of the time lag inherent in most epidemiological investigations. Each disease has its own latency period between exposure to the pathogen and the appearance of typical symptoms. Perhaps the best recent example of this is the HI virus, which is a lentivirus within the retrovirus group. These pathogens encountered in both humans and animals typically have an exceptionally long incubation time during which the individual is free of symptoms or has only very mild symptoms, a time which in animals can cover the whole life-span. Thus the mean symptom-free incubation time following HIV infection in Finland, for example, is well over 10 years. The majority of the world's HIV patients nevertheless enter the statistics only at the final stage in the infection, when they are diagnosed as having AIDS. Thus current findings regarding the areal distribution of AIDS in fact provide information on the true epidemiological situation an average of something over 10 years ago (for example, HIV/AIDS Surveillance in Europe, 1995).

Considering this time lag in the light of the changes in place of residence that subjects may have made in the intervening years, it will be appreciated that the achievement of any reliable results in epidemiological research without the use of accurate personal histories that include migration data is extremely difficult. In the case of many cancers, for instance, several decades may elapse between exposure and the development of the disease. Thus geographical analyses of the areal distribution of a particular type of cancer can easily lead to erroneous conclusions and aetiological hypotheses without personal exposure details. The problem can be avoided, of course, by obtaining research material based on spatially and temporally accurate case histories, but the compiling of such a body of data at a statistically reliable level for each areal unit would be both laborious and costly in most cases. This chapter will consider the question of what opportunities are offered by GIS for solving the problem of time–space lag in geographically oriented epidemiological research, using data available in Finland.

7.2 Time geography

The further one advances in geographical research into the spatial regularities observed in the diffusion of innovations, the more clearly one realises that this diffusion is just one part of the complex process of innovation in human culture which begins with discoveries that are located in both space and time and extends to the consequences of their diffusion. It was realisation of the complexity of the innovation process that led geographical research to address problems of time geography – with Torsten Hägerstrand (1970a, b) leading the way (Pred, 1977) as he had done in diffusion studies.

Time geography is concerned with the tracing of human activity in a three-dimensional manner, with two axes specifying location on the earth's surface and the third a time axis (Figure 7.1). By adjusting our scale on the third dimension, we can examine changes connected with daily rhythms, weekly rhythms, seasonal variations or whole human life-spans on such time–space coordinates. Then, by studying the behaviour, in space and time, of either an individual person or groups, it is possible to develop models or make generalisations or comparisons.

The main principle of this approach is that space and time form an inseparable combination in all human activities. Every event can simultaneously have a place defined for it on the spatial level (on the earth's surface, for example with respect to a source of infection or exposure to a pathogen) and on the temporal level (for example the duration of exposure). The life of every individual can be depicted as a continuous movement in time and space in which the main changes are concerned with physical structures (what Hagerstrand calls 'stations'), clusters of activities or situations formed by a plurality of individuals, or situations in which groups of this kind break up. The break-up of such activity clusters leads to the rise of new ones or causes the individuals to seek out new stations.

The second principle of time geography is concerned with conscious or unconscious constraints on the freedom of movement of the individual from one station or activity cluster to another. The first category of these is *capability constraints*, determined by human biological characteristics and the (technological) relation which

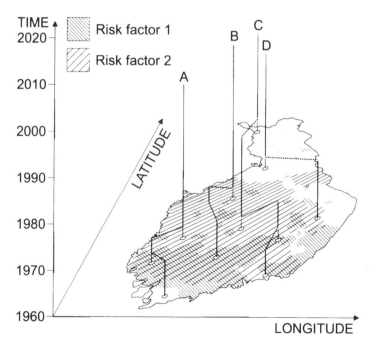

Figure 7.1 A schematic example of four individuals' life times depicted in a time–space diagram over a map of Finland. The time axis can depict any amount of time depending on the subject (cycle) of the study, such as 24 hours, a week, a year, or life time of an individual. An example of possible exposure to a health risk factor in a 24-hour cycle could be noise in a working environment. An example of exposure to a health risk factor in a life long time period could be time spent living in the vicinity of an incinerator.

each human society develops between its culture and the environment and the characterisations which it applies to this relation. For example, human beings have to obtain sufficient rest in the course of their daily rhythm to maintain their functional capabilities. A person can be in only one place at a time, and most of us are capable of executing only one task at a time. Correspondingly, the technology produced by human culture determines fairly precisely how far a person can move in the course of one day. Applied to our modern information society, it is the latter that partly explains the current discussion of distance working; there are many inhabitants of the larger metropolises who already have to spend a large proportion of their daily time budget on travelling to and from work.

The second category comprises *coupling constraints*, which determine the station of each individual with respect to various groups. These constraints enable the formation of activity clusters that serve the interests of consumption, production, social intercourse or some other purpose that can only be achieved in a group. Working communities composed of a number of individuals form a coupling constraint of this kind, as does a school for those of compulsory school age, or national service in countries whose defence relies on a system of conscription with repeated compulsory military training.

The third category is *authority constraints*. The use of space in our modern society is carefully regulated and highly selective, as all spaces are for practical purposes of finite capacity. Authority constraints are implemented through laws, statutes, regulations, economic barriers and power relations, etc., in order to create a whole body of modalities that effectively restrict the use of given areas. Although authority constraints form a complicated regulatory system in all industrialised countries, the differences between countries in this respect can be considerable.

There is one further fundamental concept connected with time geography which is of particular significance from the point of view of the geography of health, the *domain*. This is closely allied to authority constraints, as the majority of domains are determined by these. The best example of the most everyday of all domains for the individual is the home, which is usually defined quite straightforwardly in terms of ownership or occupancy rights. The prerogatives associated with the home and the legislation which regulates the privacy that the individual enjoys in his or her own home are all authority constraints aimed at safeguarding the citizen's domain. But domain can also be understood in a more complex manner, as an abstraction. Access to privately owned land is in most countries carefully restricted by various authority constraints, and in most cases trespassing is precisely defined by law, although again there may be quite considerable differences between cultures. The Nordic Countries, for example, have from time immemorial recognised a common right to enter private land and to pick berries there and use the area for other comparable purposes, and this right is also protected by law. Similarly the right to pursue a certain occupation may be regarded in a sense as an abstract domain. To function as a physician, for example, one is obliged to go through a certain accepted form of training, to register with the responsible authorities and to submit to their control, and it is only compliance with these conditions that can enable the individual, through exercise of the profession of physician, to enjoy the various prerogatives that belong to that profession.

In terms of medical geography, a domain can be almost any defined and delimited area whatsoever, either physical or abstract. It could even be determined by a feature such as pollution of the environment (for example the area affected by fallout from Chernobyl), the area to which a certain infectious disease is endemic or an area in which natural or artificial radon radiation is encountered. A domain may also form temporally restricted areas, the existence of which covers only part of the human life-span, for example an area where those who lived there at a certain time have been exposed to pollution but those moving there later have not, or at least not to the extent of it being a health hazard.

An analysis in terms of time geography, whether of a certain individual, a certain family, the elderly population or some other definable group, is always grounded in the notion that the location can always be defined by time–space coordinates. This has been illustrated by mapping the area of interest and the period of time concerned onto a three-dimensional time–space prism. Here a person who stays in one place may be represented by a vertical line (a process on the time dimension alone) and changes of place by horizontal lines (a process on the space dimension as well). Time periods usually prove to contain innumerable moves on the space dimension, which give rise to trajectories, and it is by analysing these at either the individual or the group level that we can determine people's behaviour in terms of time geography. The compilation of data by age group, for instance (for example schoolchildren, working-age population and elderly population), or by socioeconomic class or sex can open the way to more diversified treatments. An analysis in terms of gender would be essential, for example, when studying exposure to the risk of violence when moving about in a city at night.

The approach advocated by time geography spread rapidly within human geography and gave rise to a broad tradition of research extending from the level of the individual to analyses of the time–space budgets of large populations (for example Carlstein *et al.*, 1978a, b, c; Pred, 1981). The majority of this work took place in the 1970s and 1980s, however, and after the initial enthusiasm, time geography settled down to occupy an established but perhaps too modest position relative to its real importance, as one specialised branch of human geography. On the other hand, it appears to have found a role as a link between the behavioural sciences, regional planning and geography, and the most recent research may be seen as an attempt to respond to the structural change taking place in the industrialised countries and the challenges posed by the information society, the network economy and flexible manufacturing. In one of the most interesting discussions on this topic Adams (1995) considers the increasingly dynamic nature of the modern citizen's time–space budget. Adams proposes various 'extensions' to the time–space prism which represent systems based on electronic media and real-time communication. Relative to the original time–space prism, people are nowadays able to form activity clusters and groups regardless of spatial location. Video conferencing, electronic mail and even the 'traditional' television set form electronic extensions based on unidirectional or interactive communication that serve to expand the individual's functional possibilities.

7.3 Time geography and GIS

In view of the great popularity once enjoyed by research into time geography, it is somewhat surprising that only a few papers have been published in which the subject has been approached via GIS. There are two reasons for this. First, the peak activity in time geography was in the 1970s and early 1980s, when modern GIS applications were still in their experimental stage. Second, GIS as a tool for geographical research has been very much directed towards the analysis of problems existing on the spatial dimension, to the virtual exclusion of the temporal dimension. Although much effort has gone into the development of GIS software over the last ten years or so, its capabilities for temporal analysis are still very limited. Research into time geography is nevertheless in principle committed to working on both dimensions, spatial and temporal, and to regarding these as inseparable.

We frequently forget when applying GIS techniques to complex spatial problems that almost all phenomena of interest to human geography have a temporal component, and it is for this reason that the broad-based exploitation of time–space models in GIS software can be regarded as having remained underdeveloped (Peuquet, 1994). Computer cartography applications are usually atemporal, relying on the use of dots, lines, shading and other conventional techniques. As soon as we add the temporal aspect, this traditional three-dimensional (x and y for coordinates, z for magnitude) approach becomes a four-dimensional (x, y, z, and t for time) one, but at the same time it exposes the weaknesses in the presentational capabilities of our processing systems (Peuquet, 1994). The computer-assisted cartographical techniques that are available at present are founded upon three methods of expressing temporal change. One can construct a series of separate, consecutive maps in which the progression in the phenomenon being studied can be followed in the manner of a strip cartoon, although the maps seldom illustrate the situation at regular intervals in time. A second method is to construct a constant base map on which the phenomenon to be studied can be projected serially (for example for weather forecasting), and a third is to add graphs or cartograms to a standard base map in order to depict changes in the situation in different parts of the area concerned (Peuquet and Duan, 1995). All these techniques are relatively clumsy and are scarcely suitable for the description and analysis of spatio-temporal processes (see also Monmonier, 1990).

Some papers have been published in the last few years, however, in which suggestions have been made for exploiting the improved capabilities of GIS software and the increased power and capacity of computers for incorporating the time dimension. When using grid (raster) data – one of the two main methods of storing geographically referenced data – each sequential grid represents the situations at a certain moment in time. Such an approach is conceptually straightforward and any data can easily be retrieved (Peuquet, 1994). The use of grid data also provides the user with easily implemented tools to compare different data layers. Although all major GIS packages include a variety of commands and functions for grid analysis, little development has taken place as regards the design, implementation, and analysis of temporal processes using grids (Artimo, 1996; Artimo and Erke, 1996).

The other main representational approach of storing geographically referenced data is the vector model. Several papers have been published extending the vector approach to incorporate temporal dynamics into GIS (Hazelton, 1991; Kelmelis, 1991, cit. Peuquet, 1994; Langran, 1991, also 1989). All suggested extensions are based on the concept of amendment vectors, that is changes in an initial location and the time of its occurrence are recorded as additions to the vector thus maintaining the temporal continuity of each object. According to Peuquet (1994), storage of amendment vectors in practice becomes unwieldy when individual objects evolve over time, altering the topology of the connected vectors. The problem is compounded when new objects of varying kind emerge, disappear, and reappear.

Thus, the problem of representing spatio-temporal processes in GIS has been regarded as merely a question of implementation (Peuquet, 1994). Adding time into grid models or vector models leads to the design of four-dimensional representations in which the time dimension has an equal coordinate value in the cube-like space (cf. Galison, 1985). Peuquet (1994), however, evaluates such homogenous four-dimensional representations as insufficient for use in GIS because time and space exhibit important differences in their referential bases for potential queries. Instead, Peuquet (1994) presents a new approach that incorporates distinct location-, time-, and object-based components into GIS. The new approach is based on expanding the traditional dual spatial representational framework (what, where) into a triad framework (what, where, when). All information is strored so that locations, times, and objects form an interrelated scheme of the spatio-temporal process (data) studied.

The triad approach offers three advantages for process modelling. First, process models are expressed as systems of rules and constraints that represent alternative, non-deterministic states allowing complex process models to be more explicitly and robustly represented than with purely mathematical methods. Second, the triad framework accommodates analyses of complex spatio-temporal distributions with simultaneous events. Third, the framework provides efficient tools for uncovering potential spatio-temporal patterns or associations and understanding them through modelling. Peuquet and Duan (1995) provide the first practical example of the triad approach, presenting a new event-based spatio-temporal data model (ESTDM) based on time as the organisational basis: it is intended to facilitate analysis of temporal relationships and patterns of change through time. They give three algorithms for implementing an efficient and straightforward way of making temporally-based queries of geographically referenced data. For a thorough review of the subject and most recent developments, see Peuquet (1997).

7.4 Time geography and health

One of time geography's objectives is to analyse the consequences of coexistence in time and space or the effects of different phenomena existing side by side (Åquist, 1992). Although the epidemiological relevance of such an ambition can easily be perceived, the logic and visualisation technique of time geography have mostly been used to display and analyse time and space as resources in a problem-oriented way. At best, time geographical analysis has been used to identify problems in accessibility, health care provision, and in utilisation of places in situations of supply and demand (Schærström, 1996).

According to Schærström (1996), time geography has rarely, if ever, been applied or even attempted in epidemiological contexts. Despite its obvious potential it has seldom served as a tool for analysing how places might affect the health of people subject to exposure of pathogenic agents of the area. Lenntorp (1992) has shown how time geography can be used in improving data concerning past medical history for mental disorders as regards patients presenting suicidal behaviour. He concludes that there are weaknesses in the way that life events are registered and treated, and that in most cases the spatial variable is absent – despite how central its role may be in the aetiology of, for example, the presenting symptoms.

Schærström (1996) has developed Lenntorp's observations further, seeking to build a more general model of time geography's concepts and components for medical geography. It seems plausible to extend the ambitions beyond mental disorders to other kinds of medical conditions by allowing social, physical, or biological surroundings to act as potential factors in the aetiology of both mental and somatic disorders. The idea is further supported by recent findings of what is currently called an environmental disease – something that only recently has been accepted as a 'real' disorder by the medical realm.

Schærström (1996, pp. 103–105) identifies five hypothetical situations of how exposure can take place in the time–space prism. In the first case, exposure materialises when two persons meet occasionally; one is the carrier and the other susceptible. With no indication of the temporal or the spatial scales such a simplified situation does not convey any impression of the duration of exposure, and is defined as occasional and recurrent contact.

The second case is defined as continuous or intermittent exposure. Environmental factors such as air and water provide examples of this. Both elements are biologically vital to human life and often unavoidable even when contaminated. An example of continuous exposure to pathogenic agents through air is living in a large metropolitan area with thousands of sources contaminating the air.

The third case is defined as protracted exposure with delayed effect. An individual may have been exposed to a pathogenic agent prior to moving to the current area of residence. An early exposure, for example, to asbestos or radon may not develop into noticeable symptoms until decades later. As identified in the beginning of this chapter, it is the protracted exposure with delayed effect which is the most difficult problem to handle in many medical geography studies.

The fourth and fifth cases are defined according to the mobility of the source of the pathogenic agent or the susceptibles. An incinerator is a good example of an immobile source with mobile agents. Depending on the changing meteorological conditions, pathogenic agents can travel long distances reaching places and people in various parts of the world. The fifth case, immobile source with mobile susceptibles, is best illustrated by bearing in mind how varied and complex patterns different forms of human interaction produce. Daily commuting, long distance travelling and migration involve billions of moves every year all over the world.

Thinking of the multifactoral aetiology of many diseases, it may be reasonable to add a sixth hypothetical situation to Schærström's classification. This is the case of multiple exposure often required to produce certain conditions. Ultraviolet radiation is an example of this. Recent findings in biomedical research refer to the key role of ultraviolet radiation in the aetiology of many severe health disorders including not only several malignant neoplasms (Bentham, 1993) but also the onset of some viral infections such as HIV infection. Thus, we can define the sixth case as consecutive or simultaneous exposure to several agents, none of which alone may be sufficient to cause a disease.

7.5 An example: combining environmental data and individual histories

Despite the limits of currently available GIS software in handling the time dimension in analysing spatio-temporal processes, GIS already has the potential of offering a versatile set of tools for medical geography research. For the purpose of illustrating how GIS could be used for medical geography analysis, consider a hypothetical example of a research plan exploiting only currently available tools. Computer-based databases available in Finland – and in most Nordic countries as well – concerning the state of the environment and individuals are very extensive. The hypothetical model shows how we can combine environmental data and register data on individuals for obtaining data on the level and duration of exposure to specific pathogenic agents on an individual level – and gain considerable advantages in time and money spent on sampling and compiling the data.

There are two separate monitoring systems that provide data on the state of the environment. One is designed and run by the Finnish Environmental Institute (FEI), and the other is the Finnish Centre for Radiation and Nuclear Safety (FCRNS). The computerised databases consist of hundreds of variables and include geographically referenced information on hydrology, air pollution, chemicals and land use, to mention but a few. In addition, the Environmental Data System designed and provided by FEI is based on GIS, and most databases are accessible through the Internet.

Various forms of radiation have been systematically monitored in Finland for several decades. Currently, the data are sampled by use of a network of automatic measuring stations. The network provides the means for real-time monitoring of radiation levels across the country with preset alarm limits allowing for early warning. The system is supported by mobile monitoring units capable of performing very detailed sampling and limited analysis on-site. Part of the monitoring is based on systematic sampling and analysis of provisions for different contaminating agents including radioactive particles. Once compiled, these records offer high-quality longitudinal data with geographical reference of varying detail. The hypothetical example uses two different datasets; the radioactive fall-out from Chernobyl as the example of a contamination of changing spatial and temporal aspects, and the geographical distribution of naturally occurring radon as the example of contaminations of permanent spatial and temporal aspect.

After the explosion of one of the nuclear reactors at Chernobyl, Ukraine, in 1986, radioactive particles were carried north, reaching Sweden and Finland in two days. Due to changes in meteorological conditions, the majority of the fall-out was subsequently carried to central Europe, and thereafter to eastern Europe. The average kBq/m^2 level of caesium 137 was 10.7 times above normal with highest peaks reaching 73 kBq/m^2. Since the explosion, the FEI has been monitoring changes in the caesium 137 levels and its geographical distribution. Based on the database, we could design a spatio-temporal model depicting the exact level of exposure and its geographical distribution over the whole country (Chernobyl: Ten Years On: Radiological and Health Impact, 1995; Ten Years After Chernobyl, 1995).

Radon is an odourless, tasteless and invisible noble (inert) gas which is released from uranium in soil. Outdoors, radon disappears quickly. Indoors, however, radon contamination can increase quickly if proper ventilation is not arranged. Radon is considered the second most important factor causing lung cancer, after tobacco smoking. The overall exposure to radon in Finland is the highest in the world (Asikainen, 1982; Juntunen and Backman, 1991). There are several reasons for this. The climate is cold requiring heavy grounding for buildings (permafrost) and tight building routines for heating. The soils have a high uranium content and during the latest glacial epoch glacial and glacifluvial processes caused considerable erosion, sorting and deposition of debris in Finland. The geographical distribution of radon follows the major eskers as they are made of porous materials allowing more air flow than other debris. The differences in the uranium level in various soils and rocks also vary considerably adding to the geographical variation of the radon

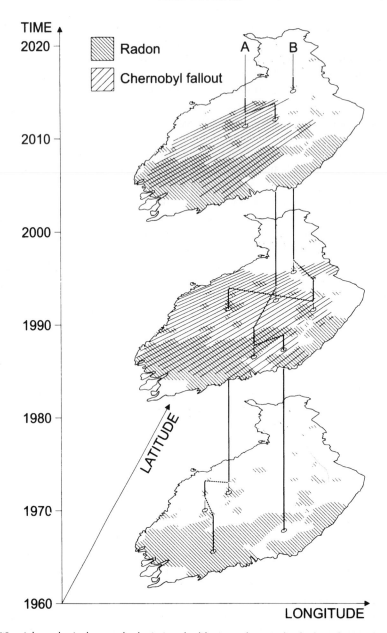

Figure 7.2 A hypothetical example depicting the life time of two individuals with varied exposure to radon gas and Chernobyl fallout in Finland. Radon, released from Uranium in soil, has a distinct yet limited geographical distribution with very little longitudinal variation in Finland. Chernobyl fallout has a less clear-cut geographical distribution and a longitudinally changing character of diminishing radiation.

problem in Finland. As a result, the radon problem is mainly located in south and south-east Finland. Data on radon are available by municipality, that is by over 450 small-area units, and by coordinates if interpolated data are accepted.

All administrative records in Finland – changes of address, taxation, family status, dwelling, to mention but a few – are all registered on a personal level and spatially referenced using a unique identification code for each individual (social security number). A system of keeping detailed personal registers has a history of several hundred years and is well-respected by both authorities and the public. All registers are currently computerised and all information is linkable on an individual level. The word 'census' in Finland simply means a number of computer runs which will depict the status at a certain moment in time. Provided it is done for legitimate purposes and obeying very strict data disclosure rules, such data could be used in a time geographical analysis in various ways.

Let us consider a research plan aiming at establishing how many lung cancers are caused by radon in Finland. It is obvious that we wish to sample a geographical and statistically significant sample of people living in areas where radon exposure is known to be above the average level or higher. Also, we will need information on a matched control group of people living in areas with less than average or no radon exposure at all. We also need to sample data on different confounding variables such as socioeconomic data, smoking data, nutrition patterns and occupational risks (for example exposure to asbestos) on each individual. One of the confounding factors we also want to control for is the contamination caused by the fall-out from Chernobyl. Although we may assume that the radioactive components reaching Finland (mainly caesium 137, but also zirconium 95 and rutenium 106) are not known as causative agents of lung cancer, it would be necessary to control this factor carefully.

Using GIS, we could design a detailed spatio-temporal database with downloaded information on radon level and caesium 137 levels, and their spatio-temporal variation by small-area unit (for example municipality) or even more detailed if necessary. Data on other variables needed for the study could also be downloaded from Statistics Finland or from other organisations keeping the records. Data on cancer incidence could be obtained from the Finnish Cancer Registry which has individual level records of all histologically confirmed cancer cases diagnosed in Finland. Every step of designing and compiling this database can be done automatically by use of the computer-based databases. Using individual migration histories we could produce a detailed spatio-temporal case history of the level and duration of exposure to each suggested causative agent under scrutiny. This could be done by running individual migration histories through the time–space prism depicting high risk and low risk areas (Figure 7.2). Having produced the variables measuring the exposure, we could then proceed to analysing the data using the statistical and epidemiological methods deemed necessary for producing the results.

Without considering the technicalities (for example how to design the database, how to download the data, or how to write the macros), we can conclude that compared with more traditional ways of compiling the data (postal questionnaires, personal interviews, manual matching of case histories with environmental data), the use of GIS and environmental data and individual case histories arranged in the form of a time–space prism is faster, more cost-effective and more versatile.

7.6 Concluding remarks

The tools offered by currently available GIS software have some potential for performing analysis on health-related spatio-temporal processes. Although the hypothetical example explained above is only a moderate attempt to exemplify this, combining environmental data with individual level register data provides an illustrative example of the many uses of GIS in a framework offered by time geography. We could easily employ more complicated research designs such as individual level time–space databases with very detailed geographical referencing to model and forecast the simultaneous spatio-temporal diffusion of a contagious pathogenic agent. Also, the use of molecular epidemiology with help from recently developed phylogenetic methods may open completely new worlds when connected with the framework put forward by time geography.

In most geographical research the spatial and temporal dimensions are inseparable. The key to understanding many of the elementary geographical concepts is based on the identification, description, and analysis of spatio-temporal processes. This is also the case when aiming at understanding most health-related questions. Currently evolving trends of developing GIS software to offer more versatile and powerful tools for handling spatio-temporal processes will eventually provide us with new opportunities for medical geography research and other health-related inquiries.

Until very recently, GIS software has evolved as a powerful set of tools with little emphasis on temporal dimension. The extensive theoretical review provided by Peuquet (1994) and the subsequent work on designing temporally oriented GIS applications (Peuquet and Duan, 1995) underline the potential offered by GIS in the description and analysis of spatio-temporal processes. It now seems that an increase is taking place in genuinely time-oriented GIS research in the field of time geography and that we shall find that future GIS software possesses combined spatio-temporal analytical tools. It is reasonable to say that further development of, for example the triad approach will open new ways for the use of GIS in medical geography research.

But there are other lines of development which we should consider when designing new GIS-based approaches to health-related research with temporal dimension in mind. Perhaps one of the most challenging is incorporating different concepts of space into GIS. In a vast majority of research done using GIS, the world is understood and implemented in terms of absolute space. Most decision-making on all levels of human activity, however, takes place in a relative space. Since decisions taken by individuals can have long-prevailing effects on their health, it may be of importance to analyse such decisions in other than absolute space. And when it comes to analysing different forms of power relations known to affect the health status of an individual, we need to understand the world in terms of relational space (Harvey, 1973). Seeking for even more challenging research situations in which time–space prisms could be used as a framework, it is evident that multilevel-level modelling needs to be considered (for example Jones and Duncan, 1995). GIS has the potential of putting time and space together – especially when the computing capacity is increasing and new algorithms are designed – but it will also require fundamental development at a theoretical level.

References

ADAMS P. C. (1995) A reconsideration of personal boundaries in space-time. *Annals of the Association of American Geographers*, **85**, 267–285.

ÅQUIST A.-C. (1992) Tidsgeografi i samspel med samhällsteori. *Meddelanden från Lunds Universitets Geografiska Institutioner*, Avhandlingar, 115, Lund University Press, Lund.

ARTIMO K. (1996) Paikkatietojärjestelmät ympäristövaikutusten arvioinnissa – erityisesti maankäytön suunnittelussa. *Ympäristösuunnittelun uudet tuulet – GIS paikkatietopäivä*, MTT, Jokioinen, pp. 14–17.

ARTIMO K. and ERKE M. (1996) Use of grid-based grid-analysis and processing in land use planning. *Abstracts of the ISPRS Conference*, Vienna, 1996.

ASIKAINEN M. (1982) Natural radioactivity of ground and drinking water in Finland. *Finnish Centre for Radiation and Nuclear Safety*, STL-A39.

BENTHAM G. (1988) Migration and morbidity: implication for geographical studies of disease. *Social Science and Medicine*, **26**, 49–54.

BENTHAM G. (1993) Depletion of the ozone layer: consequencies for non-infectious human diseases. *Parasitology*, **106** (Suppl.), 39–46.

BERNARDINELLI C., CLAYTON D., PASCUTTO C., MONTOMOLI C., GHISLANDI M. and SONGINI M. (1995) Bayesian analysis of space-time variation in disease risk. *Statistics in Medicine*, **14**, 2433–2443.

CARLSTEIN T., PARKES D. and THRIFT N. (eds) (1978a) *Making Sense of Time*. Edward Arnold, London.

CARLSTEIN T., PARKES D. and THRIFT N. (eds) (1978b) *Human Activity and Time Geography*. Edward Arnold, London.

CARLSTEIN T., PARKES D. and THRIFT N. (eds) (1978c) *Time and Regional Dynamics*. Edward Arnold, London.

Chernobyl: Ten Years On: Radiological and Health Impact (1995) OECD Nuclear Energy Agency.

GALISON P. L. (1985) Minkowski's space-time: from visual thinking to the absolute world. *Historical Studies in the Physical Sciences*, **10**, 85–121.

HÄGERSTRAND T. (1970a) Tidsanvändning och omgivningstruktur. Urbanisering i Sverige: en geografisk samhällsanalys. *Statens Offentliga Utredningar*, 1970/14, Stockholm.

HÄGERSTRAND T. (1970b) What about people in regional science? *Papers of the Regional Science Association*, **14**, 7.

HARVEY D. (1973) *Social Justice and the City*. Blackwell, Oxford.

HAZELTON N. W. J. (1991) Integrating time, modelling and geographical information systems: development of four-dimensional GIS. Unpublished PhD, Department of Geography, Pennsylvania State University.

HIV/AIDS Surveillance in Europe (1995) Quarterly Report 46. European Center for the Epidemiological Monitoring of AIDS.

JONES K. and DUNCAN C. (1995) Individuals and their ecologies: analysing the geography of chronic illness within a multilevel modelling framework. *Journal of Health and Place*, **1**, 27–40.

JUNTUNEN R. and BACKMAN B. (1991) Radiogenic elements in Finnish soils and groundwaters. *Applied Geochemistry*, **6**, 169–183.

KELMELIS J. (1991) Time and space in geographic information: toward a four-dimensional spatio-temporal data model. Unpublished PhD, Department of Geography, Pennsylvania State University.

LANGRAN G. (1989) A review of temporal database research and its use in GIS applications. *International Journal of Geographical Information Systems*, **3**, 215–232.

LANGRAN G. (1991) *Time in Geographic Information Systems*. Taylor & Francis, London.

LENNTORP B. (1992) Biografier, diagnoser och prognoser. In *Självmord som existentiellt problem*, Folksams vetenskapliga råd, Fårskningsrådsnämnden Rapport, **92**, 63–76.

MILLER H. J. (1991) Modelling accessibility using space-time prism concepts within geographical information systems. *International Journal of Geographical Information Systems*, **5**(3), 287–301.

MONMONIER M. (1990) Strategies for the visualization of geographic time-series data. *Cartographica*, **27**(1), 30–35.

PEUQUET D. J. (1994) It's about time: a conceptual framework for the representation of temporal dynamics in geographic information systems. *Annals of the Association of American Geographers*, **84**, 441–461.

PEUQUET D. J. (1997) Time in GIS and geographic databases. In Longley, P., Maguire, D., Goodchild, M. F. and Rhind, D. W. (eds) *Geographical Information Systems: Principles and Applications*, Longman, Harlow (forthcoming).

PEUQUET D. J. and DUAN N. (1995) An event-based spatiotemporal data model. ESTDM for temporal analysis of geographical data. *International Journal of Geographical Information Systems*, **9**, 7–24.

PRED A. J. (1977) The choreography of existence: comments on Hägerstrand's time-geography and its usefulness. *Economic Geography*, **53**, 207–221.

PRED A. J. (ed.) (1981) *Space and Time in Geography: Essays Dedicated to Torsten Hägerstrand*. Lund University Press, Lund.

SCHÆRSTRÖM A. (1996) Pathogenic paths? A time geographical approach in medical geography. *Meddelanden från Lunds Universitets Geografiska Institutioner, Avhandlingar*, 125, Lund University Press, Lund.

Ten Years After Chernobyl (1995) Finnish Centre for Radiation and Nuclear Safety, Helsinki.

PART TWO

Applications

APPENDIX

CHAPTER EIGHT

GIS Applications for Environment and Health in Italy

STEFANIA TRINCA

8.1 Introduction

Knowledge of the geographical distribution of environmental phenomena and factors is extremely useful in the study of disease causation, in the design of prevention strategies and in public health in general. Given the need to access relevant information, interest is growing in tools for the storage, management, processing and display of different types of spatial data. As a result, database management systems (DBMS), packages for statistical analysis, software for mathematical modelling, image processing systems, telematic facilities and systems for data transmission all have a role to play. However, the linking of different types of data, such as those regarding population (for example census and socioeconomic data), health (for example mortality, morbidity data), and environment (for example monitored data on pollution) is possible with Geographical Information Systems (GIS), where, ideally, such diverse tools and systems may be integrated.

In Italy, GIS have already been introduced to manage those aspects of infrastructure where locations are generally delineated quite precisely; so applications in areas such as urban planning, the utilities (water, electricity and gas industries), and transport planning are now quite common. The application of GIS to the health field is quite recent, but interest is growing among those involved in environmental and health research and some examples in the fields of geographical and environmental epidemiology, risk assessment, and public health are examined below. Since 1990 an ever-increasing number of managers and researchers in the National Health Service have attended courses organised by the Data Management Service of the Istituto Superiore di Sanità (ISS), the Italian National Health Institute, whose prime objective is to foster an interdisciplinary approach to the study of environmental and health related issues, and to present the latest available instruments that may be used to this purpose. In this context, GIS have become fundamental.

8.2 GIS applications in the environmental field

We consider first some applications in the environmental arena. This was one of the first to benefit from the introduction of GIS, since such tools are highly suited for the description and analysis of physical and environmental phenomena.

The Italian public bodies that are responsible for the protection of the environment at the regional, provincial and municipal levels are engaged in the creation of Local Information Systems (LIS) involving GIS, some of which are devoted to the management and control of environmental factors of utmost importance for public health (such as air, water and soil pollution). Examples (Figure 8.1) are the system for air pollution data management implemented by the Assessorato per l'Energia e l'Ambiente of the Lombardia Region (Regione Lombardia, 1994), and the system for surface water control of the Provincia Autonoma di Trento (Provincia Autonoma di Trento, 1996). Both these LIS have been running for a few years now and are valid instruments for the management and the control of the environment. They are capable of yielding useful data to set up GIS applications for the study of the relationship between environment and health.

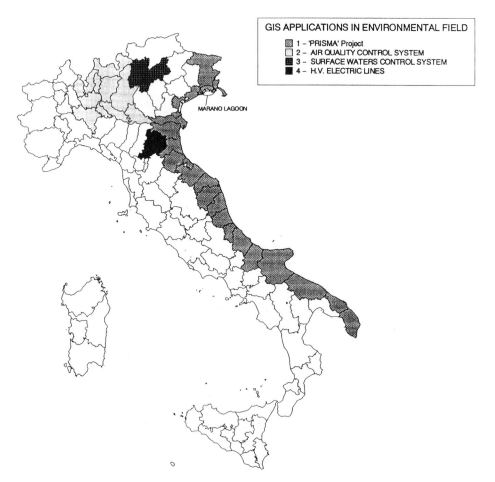

Figure 8.1 Geographical distribution of the described GIS applications in the environmental field.

At the local level it is possible to find systems devoted to the control of specific risk factors. For example, the Province of Bologna has mapped the distribution of high-voltage electric lines and linked it with urban development data. From the viewpoint of town planning, this clearly represents an instrument for decision making, but given concerns expressed about the possible associations between exposure to electromagnetic fields and human ill health it could also be considered a powerful means for epidemiological monitoring. Elsewhere, in the Friuli Region, the local health unit of Bassa Friulana has developed a GIS application to monitor the quality of water in the Marano lagoon. This is a particularly interesting area with summer resorts on the Adriatic coast, and mollusc breeding farms, and the monitoring of water quality for its possible impact on public health is clearly crucial.

At a more regional level on the Adriatic, the Italian Ministry for the University and Scientific and Technological Research, in close collaboration with the major scientific Italian institutions, is developing the PRISMA2 Project (Programma di Ricerca e Sperimentazione per il Mare Adriatico) with the aim of carrying out research to identify and define the necessary actions for the safeguard of the sea and coastal areas (Figure 8.1). PRISMA2 is a multidisciplinary project involving several research teams that work in oceanography, marine biology, ecology, toxicology, and epidemiology. The Istituto Superiore di Sanità will be responsible for the health aspects of the project, including the health of the population, the quality of coastal waters, and the control of sea food produced in the area. The information system sustaining the project will involve the use of GIS and therefore all the collected data will have to be georeferenced.

This brief review indicates that GIS applications in environmental monitoring relevant to human health are prominent, especially in northern Italy (Figure 8.1). It should be kept in mind that, since there is currently no integrated nationwide system to collect environmental data, their availability varies widely within the country and is restricted to a few specific environmental domains. It is therefore of utmost importance to establish efficient systems for the exchange of information, firstly among the people who work in the environmental field and secondly for those who intend to use data on the environment in order to deal with other issues directly or indirectly related to it.

8.3 GIS applications in the health field

Health data are intrinsically difficult to describe spatially, and their association with purely environmental phenomena is not always straightforward because there are often several factors leading to a specific health outcome. This is one major reason for the delay in the introduction of GIS to public health. As interest in this tool increases in Italy, applications are being developed but only a few have yielded concrete results so far.

8.3.1 Epidemiology

In Italy, the applications of GIS to epidemiological research deal essentially with mortality data. Being part of routine official statistics, these data are, in general, coherent and consistent, though they lack high-resolution spatial detail. Routine data are only available down to the level of the municipality, and access is restricted to authorised institutions for well-documented research reasons.

The Institute for Medical Statistics of the University of Milan, in collaboration with the Emilia-Romagna Region and the Italian Ministry of Health, has set up a geographical information system using mortality data, one output of which is The Italian Atlas of Mortality (Cislaghi *et al.*, 1995; see also chapter 9). This describes spatial variation among small areas using kernel estimation and might, with care, be used to identify high-risk areas. Thirty-one causes of death are analysed. The tables, which cover the entire Italian territory, refer to regions that are circular in shape, with a radius of approximately 100 kilometres, and which contain a variable number of municipalities; each municipality may appear in more than one circle. Available in electronic form to all health authorities that ask for it, it is a very good starting point for environment–health correlation studies.

In 1986 a Register of Causes of Death was set up in Lazio, the region which includes the city of Rome. The Register is managed by the Epidemiologic Unit of the Lazio Region which has recently developed GEO.S.I.M., a geographical information system on mortality. This system, first released in April 1996, uses mortality data for Rome and Lazio between 1987 and 1994. It uses two levels of territorial aggregation: first, census enumeration districts (about 6000) within Rome; and, second, municipalities outside Rome (about 400 in the whole region). The system also includes annual demographic statistics and the data from the 1991 census. Enumeration districts were assigned a socioeconomic index by applying factor analysis to a subset of variables from the 1991 census. Using this index the confounding effect of socioeconomic conditions can be controlled when studying Standardised Mortality Ratios (SMRs) for selected causes of death.

GEO.S.I.M. therefore describes cause-specific mortality at small area levels. The possibility of analysing mortality data at a small geographic scale, and the availability of data for areas that are reasonably homogenous as far as demographic characteristics and environmental exposure are concerned, offers a starting point for more specific analyses. Smoothing procedures are used for the estimation of SMRs. These procedures minimise the random fluctuations of estimates and yield maps that admit a more robust epidemiological interpretation. The analysis is performed using the BEAM statistical software that uses a Bayesian approach in estimating SMRs (see chapter 10). Two maps (Figure 8.2) illustrate the possibilities, using data on mortality due to lung cancer in males in the Lazio Region. The maps represent the traditional (Figure 8.2a) and the Bayesian (Figure 8.2b) estimation of SMR. Their comparison clearly shows the advantages of the latter. In the first, Poisson variability is a feature of areas where a small number of cases occurred. In the Bayesian map, instead, the area at the highest risk is either the result of a particularly high SMR and a high number of observed events (in which case there is no need to incorporate information from adjacent areas), or it is due to the fact that the adjacent areas also present a higher risk (in which case a combined effect of these two factors cannot be excluded). In short, the Bayesian map is more reliable in the definition of areas which call for *ad hoc* surveys to identify environmental risk factors.

GEO.S.I.M. can also identify suspicious clusters of deaths for specific causes, and it allows the user to perform mortality studies near pollution sources. Stone's conditional test, further developed by Elliott *et al.* (1992), is used to analyse risk as a function of distance (Michelozzi *et al.*, 1998). The test is based on the hypothesis

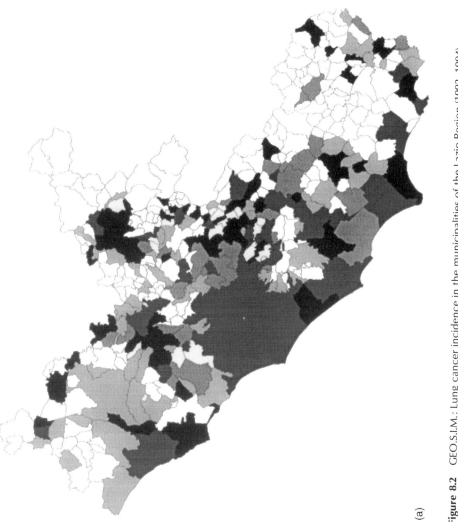
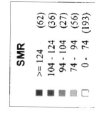

(a)

Figure 8.2 GEO.S.I.M.: Lung cancer incidence in the municipalities of the Lazio Region (1992–1994) (a) raw SMR.

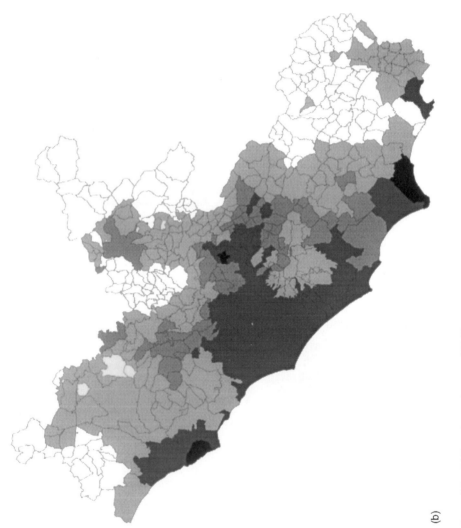

Figure 8.2 (b) Bayesian estimates of SMR.

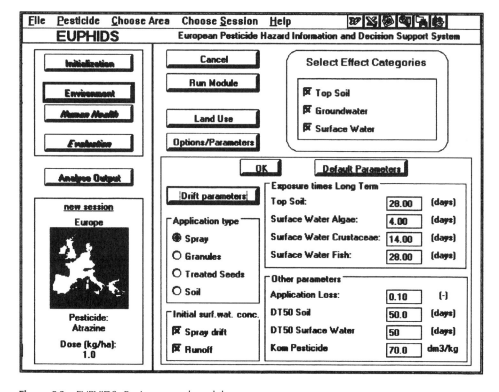

Figure 8.3 EUPHIDS: Environmental module.

that the risk of disease decreases as the distance from the source of pollution increases, and this is evaluated against a null hypothesis of uniformly distributed risk.

At the Istituto Superiore di Sanità the research team dealing with GIS are working on two applications. One, in collaboration with the Health Statistics Unit of the same Institute, is a project to create a geographical database at various spatial levels to acquire health data from current sources (mortality and morbidity registers) and integrate them with demographic and census data. The goal is to develop a system capable of giving a picture of the health situation of the country that can be exploited for epidemiological studies. The other application, in collaboration with the Laboratory of Epidemiology and Biostatistics, is devoted to an analysis of mortality from accidents, with the aim of studying the relationship between the various events (traffic, work place, home accidents) and environmental factors, including socioeconomic conditions and lifestyle.

8.3.2 Risk assessment

GIS applications for health risk assessment are hard to develop because of the lack of data on exposure levels of the population. As noted earlier, enviromental data are fragmentary and not always comparable. Morbidity data are also difficult to

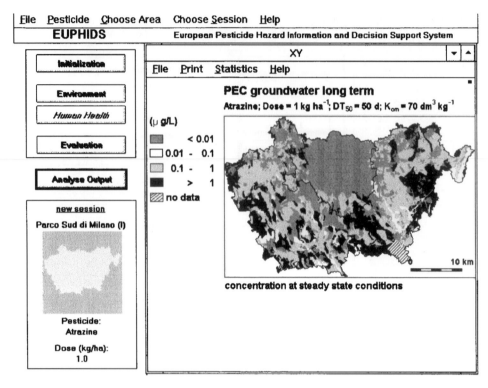

Figure 8.4 EUPHIDS: Environmental module – groundwater risk assessment evaluated in the area of Parco Sud di Milano.

obtain, especially those which are valid indicators of exposure. There are, nevertheless, some interesting examples of ongoing work.

The International Centre for Pesticide Safety (ICPS) has been created by the Government of the Lombardia Region in cooperation with the University of Milan, and with the agreement of the Italian Ministry of Health. Its mandate is to work as a technical collaborating Centre (with the WHO Regional Office for Europe) in the field of pesticide management and safety. It is working on projects aimed at developing methods for the description, analysis and study of pesticide environmental pollution, and associated effects on the population. To this end, ICPS has developed a joint project with the Rijksinstituut voor Voldsgeondhenongheid en Milieuhygiene (RIVM) in Bilthoven, The Netherlands, and the Frauhofer Institut fur Umweltchemie und Ökotoxikologie, Schmallenberg Grafschaft, Germany.

In this project, software known as EUPHIDS (European Pesticide Hazard Information and Decision Support System) has been developed. EUPHIDS offers a powerful tool to support decision makers in pesticide registration, at different spatial scales. The system adopts scientifically sound prediction models for phenomena such as leaching, run-off, spray-drift, exposure of the operators and exposure of the general population through diet. For each of these phenomena, algorithms have been developed to estimate the dose delivered to the targets. Moreover, the models have been applied in such a way that the specific features of each environment can be taken into account at various scales (continental, national, local) and the spatial variation of the assessment of risk can be described with appropriate maps (Beinat and van den Berg, 1996). EUPHIDS is an open, flexible system and uses state-of-the-art knowledge (data, models, rules) to provide the decision maker with information, as set out in Annex VI of the Directive 91/414.

The approach EUPHIDS follows is that of integrating data collection, data analysis, data presentation, and evaluation procedures (Figure 8.3). The risks to people and the environment, expressed as the ratio of exposure to no-effect levels, are presented in the form of maps. In this way high-risk regions as well as the components determining the risk can easily be identified. For example, Figure 8.4 shows estimated concentration levels for the pesticide atrazine in the area Parco Sud di Milano, while at the national scale Figure 8.5 gives estimates of the predicted maximum daily intake of lindane.

ICPS has also designed a project for the Lombardia Region, part of the Regional Plan, for the environmental and health surveillance of pesticide effects between 1994 and 1996. The project includes the creation of a database at various geographic scales, incorporating data on pesticide pollution and on the health status of the population. The environmental data include: hydrogeologic features; soil use; sources of pesticide pollution; chemical–physical and (eco)toxicological properties of pesticides; pesticide uses; and the concentration of pesticides in different environmental media and in particular residual concentrations in drinking water and food. Data for human populations include demographic statistics and data on specific health outcomes.

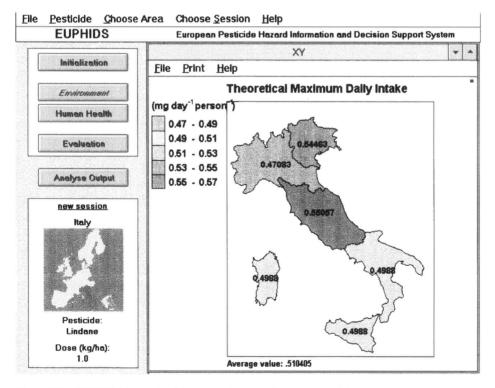

Figure 8.5 EUPHIDS: Human health – general population exposure risk assessment.

This database allows for (a) the description and representation of risk factors via thematic maps; and (b) the analysis of spatial and temporal trends of environmental and human data.

Analysis of further data on the contamination of wells in the Lombardia Region, for the period 1986–1993, has shown that contamination by pesticides of groundwater sources has decreased during the study period. The shallow aquifers of the selected areas are considered to be vulnerable to the leaching of substances; the pollution risk should be considered particularly high in areas of rice cultivation (Riparbelli *et al.*, 1996).

In a small area of Lombardia a land-use map was combined with a qualitative methodology in order to identify vulnerable areas (Azimonti *et al.*, 1994; Riparbelli *et al.*, 1995). In this way, a broad identification of potential pesticide pollution areas was first obtained. This approach was then improved by using the results of a leaching model (PESTLA) within a GIS. PESTLA manages specific information on soil, climate, culture and physico-chemical characteristics of the pesticide under evaluation, and the resulting maps represent the risk of pollution for groundwater deriving from a selected pesticide.

8.3.3 Health care management

After the creation of the National Health Service in 1983, the Local Health Units are being restructured. LHU managers and technical directors are looking for new methods and tools to assist in the management and planning of health care services and resources. In several Italian provinces some LHUs are already using GIS to develop their own information systems to link the most relevant attributes (services, population, health outcomes and socioeconomic data).

8.4 Conclusions

The Italian scenario shows that, even if still at an early stage, GIS applications are rapidly expanding in health and environmental studies. The potential of these tools has yet to be fully exploited and most of the examples that have been described here are essentially used for visualisation and hypothesis generation purposes.

There are several barriers to the more rapid adoption of GIS technology for environmental and health applications in Italy. One is the human constraint of the time needed to acquire specific training in new tools. Another is the set of problems regarding data availability. Digital cartographic data, especially at large scales, is both rare and costly. There is no clear policy to address the collection of health data in a fashion suitable for use in GIS applications. For example, data for environmental monitoring are scarce and inconsistent, while morbidity data (on disease incidence rather than mortality) are sparse, especially at local spatial scales. A further barrier is that some researchers still have difficulty in conceiving interdisciplinary work that integrates health, environment and society and the economy, and those who do try to link their own data to other specific sectors find that data are incomplete, hard to get hold of, and difficult to integrate when the data have been collected for other purposes.

GIS are tools that favour an interdisciplinary approach to the solution of problems; this feature should not be underestimated. It is the aim of the team currently working with GIS at the Istituto Superiore di Sanità to develop health–environment GIS applications from a multidisciplinary viewpoint, and to disseminate the use of these tools through courses and seminars for interested researchers and National Health Service professionals.

Acknowledgments

The author is grateful to Professor Giorgio Cortellessa for his invaluable help; Paola Michelozzi of the Osservatorio Epidemiologico della Regine Lazio and Giovanna Azimonti of the International Centre for Pesticide Safety, Legnano, for their help and kind permission to reproduce Figures 8.2–8.5; and Monica Brocco for the linguistic revision of the manuscript.

References

AZIMONTI G., RIPARBELLI C. and MARONI M. (1994) Sensitivity of soil to pesticide leaching: a combined application of a leaching model and a GIS, *Proceedings EGIS/MARI'94, Fifth European Conference and Exhibition on Geographical Information Systems, EGIS*, p. 1026.

BEINAT E. and VAN DEN BERG R. (1996) EUPHIDS, a decision support system for the admission of pesticides. Rijksinstituut voor Volksgezondheid en Milieuhygiene (NL), Vrije Universiteit Amsterdam (NL), Fraunhofer Institute für Umweltchemie und Okotoxicologie (D), International Centre for Pesticide Safety (I). EC contract No. EV5V-CT92-0217. Final report.

CISLAGHI C. et al. (1995) L'Atlante Italiano di Mortalità a Livello Comunale. *Epidemiologia e Prevenzione*, **19**, 132–141.

ELLIOTT P., WESTLAKE A. J., HILLS M. et al. (1992) The Small Area Health Statistics Unit: a national facility for investigating health around point sources of environmental pollution in the United Kingdom. *Journal Epidemiology and Community Health*, **46**, 345–349.

MICHELOZZI P. et al. (1998) A small area study of mortality among people living near multiple sources of air pollution. Il sistema informativo della mortalità su base geografica del Lazio. Esempi applicativi di analisi geografica su dati di routine. *Occupational and Environmental Medicine. Epidemiologia e Prevenzione* (in press).

PROVINCIA AUTONOMA DI TRENTO (1996) Qualità delle Acque Superficiali – Elaborazione Dati Qualità Anni 1991–1992–1993–1994. Trento.

REGIONE LOMBARDIA (1994) Settore Ambiente ed Energia. *Bollettino Qualità dell'Aria*. Milano.

RIPARBELLI C., FERIOLI A., AZIMONTI G., REGIDORE C., BATTIPEDE G. and MARONI M. (1995) Impact of pesticides to groundwater resources in an alluvional plain using a Geographical Information System. *Central European Journal of Public Health*, **4**, 21.

RIPARBELLI C., SCALVINI C., BERSANI M., AUTERI D., AZIMONTI G., MARONI M., SALAMANA M. and CARRERI V. (1996) Groundwater contamination from herbicides in the region of Lombardy – Italy. Period 1986–1993. *Atti del X Symposium Pesticide Chemistry–Piacenza*, 550–562.

CHAPTER NINE

A Multipurpose, Interactive Mortality Atlas of Italy

MARIO BRAGA, CESARE CISLAGHI, GIORGIO LUPPI and CAROLA TASCO

9.1 Introduction

The geographical analysis of mortality is that branch of descriptive epidemiology that has flourished for many years, leading to the publication of numerous atlases (for example, Mason *et al.*, 1975; Junyao *et al.*, 1979; Kemp *et al.*, 1985; Gardner *et al.*, 1993). Such atlases differ in a number of ways, not least of which is variation in the level of spatial aggregation at which data are mapped.

This development and prodution of such atlases is linked with the hope of finding new interpretative hypotheses by means of the visualisation and exploration of differences in the geographical distribution of mortality events and, consequently, in disentangling the contextual effect, which is due to the spatial location, from the compositional effect, due to population structure. To do so, constant efforts are made to obtain more reliable, analytical and interpretable spatial images. On the one hand there has been the slow process of improving the certifications; this has been achieved principally by an increasing awareness on the part of the certifying doctors of the epidemiological importance of their work. This has been paralleled on the other hand by an improvement in the statistical and computer-based instruments used in the spatial analysis of deaths (Marshall, 1991).

In Italy, much research effort has focused on the geographical analysis of mortality data, aided by the fact that the Italian National Statistical Institute (ISTAT) has, since 1980, constructed computer-based mortality databases, recording detailed information, such as the place of residence and birth of the deceased, even if, in fact, there is limited access to such information for reasons of confidentiality. However, the research reported here seeks to move away from the idea of a fixed set of tables and maps representing the geographical distribution of mortality rates, towards an interactive atlas for disease mapping and cluster identification. The reasons underlying the development of a flexible graphical and statistical tool were the growing need for proactive surveillance at small area level and the possibility to investigate the existence of local health problems quickly in response to specific questions. This requires that the decision of what has to be mapped, for which subgroup of people, and for what time window, has to be left to the final user of the information. This could be a very dangerous choice, due to the logical flaws of *post hoc* inquiries (unusual events are such by self-definition) and the complexity of interpreting spatial analysis based on aggregated data. Being aware of both problems, the strategy followed to reduce their impact was to identify a group of people, for each region, who had a substantial knowledge of their territory and who were interested in attending a series of courses oriented to the use and interpretation of the statistical procedures implemented in the new Italian atlas. The final course was based on group discussion with a tutor, where the groups comprised the 'experts' of adjacent regions and the aims of the group discussion were to read and interpret all the outputs produced for a specific set of neighbouring regions. This part was an important step for evaluating the degree of homogeneity of the different regional groups, for encouraging inter-regional cooperation and for identifying cross-border as well as intra-regional health problems.

In what follows, we first outline the data sources and analytical methods employed, before considering the design and implementation of the system. Examples relating to lung and stomach cancer are then presented.

9.2 Data and analytical methods

9.2.1 Data

The database included in the electronic atlas was provided by the National Statistical Institute (ISTAT), but was made available by the Planning Office of the Health Ministry in modified form in order to screen the necessary fields and aggregate the values of some variables (for example the individual age). This was done to render practically impossible the identification of individuals. This procedure was necessary in order to overcome the problems regarding the protection of citizens' rights and privacy as expressed in Italian legislation. The mortality data used here to illustrate the ideas involve the period 1981–1988, although a newer version of the electronic atlas contains mortality data updated to 1994.

It was decided to distinguish between migrant and stable populations and to compile an atlas regarding the latter. Persons resident in a province (P) and born in the same province or in a neighbouring one (a) are considered 'stable'; if born in other provinces (m) they are considered 'migrant' (Figure 9.1). Studies on migrant populations within Italy (Vigotti and Cislaghi, 1988) have shown a strong 'birth-effect' on mortality which, especially in studies of small areas, confounds the effects of residence which are precisely those which need to be highlighted in the mortality atlas. Analyses of migrant groups will be considered in later studies.

From 1981 to 1988 the deaths in the stable population represent 84.5% of the Italian deaths; the percentage varies greatly, however, from area to area and also depends on sex, age, and cause of death. Data on population is that provided by the 1981 census and it has also been stratified following the same criterion mentioned earlier when describing stable or migrant population. Finally, the spatial coordinates (latitude and longitude) of the 8100 Italian municipalities were also provided by ISTAT.

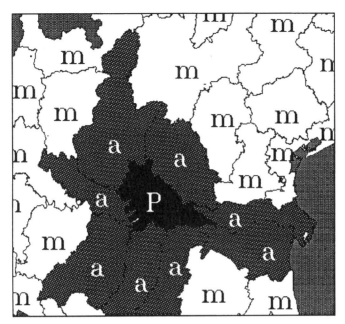

Figure 9.1 Definition of the stable population (resident and born in the province 'P' or in neighbouring ones 'a').

It is possible to choose any group whatsoever of specific cause of death. However, in order to avoid unstable relative risk estimates, the analysis at small area level cannot legitimately be performed on relatively infrequent causes. Special spatial analyses are needed in order to study such causes both at aggregate (Bernardinelli and Montomoli, 1992) and individual levels (Waller and Jacquez, 1995). This problem also involves the quality of the death certifications. In some forms of cancer, the variability due to the error of certification is greatly reduced, but this is not so for other causes, especially when dealing with older age groups.

Given that the reference population relates to the 1981 census, whilst the data on mortality refer to the period 1981–1988, there is a notable temporal difference between the two data series. When working with small areas, a municipality may have undergone notable changes in size and demographic structure. All of this can be even further emphasised in the subgroups of population, both stable and migrant, and there may be uncertainty concerning whether the information provided by the census is homogeneous with that for mortality. There are, therefore, potential problems in using data from the 1981 census as denominators for the mortality indexes. More up-to-date information regarding the population on a national level is not available. As a result, proportional indicators were used, so guaranteeing the homogeneity between the numerator and denominator of the ratios. Another reason for preferring the proportional analysis is due to the presence of some coding errors in the municipal residence of the deceased.

The values of the standardised proportional mortality rates (SPMR) do not differ that much from the values of the Standardised Mortality Rates (SMR), providing that the general mortality does not vary excessively and that the analysed cause is rare: it is due to the last reason that the SMR were calculated regarding general mortality, and the SPMR regarding single-death causes.

9.2.2 Analytical methods

One of the characteristics of this atlas is that it makes no reference to provincial or regional boundaries; the municipality boundaries are the only ones considered and even these could have been put aside had the individual records provided precise coordinates concerning the exact residence of the deceased and not just the municipality of residence (as is possible in Britain and North America, for example). The Italian territory has been represented as 31 partially overlapping circles which, by default, have a maximum radius of 100 kilometres or, whenever the number of municipalities is too large, contain at most 700 municipalities. These circles have the centre placed in selected municipalities. However, all these parameters (number of circles, circle dimension, location of the circles) may be modified by the user.

The spatial distribution of the relative risk measures, after removing the effect of known explanatory factors, is affected by two other components, which are together named extra-Poisson variability. The first component, called heterogeneity, is not spatially structured while the second, which is named clustering, is geographically organised (Breslow and Clayton, 1993; Tasco et al., 1993). In order to find the aggregate component, estimates of the density functions obtained with kernel indicators were used (Bithell, 1990; Cislaghi et al., 1995a). They can be easily interpreted as moving spatial averages of the adjacent values of each chosen point, weighted by an inverse function of distance. The resulting effect is the smoothing of the distribution with a noise reduction and a greater clarity of the signal; it goes without saying that, in this way, a distorting factor is introduced which hopefully will not alter the main image.

Present in the statistics which can be processed with this new atlas, there is – for every cause and centroid – a local analysis of the trend accompanied by the corresponding map (either drawn on screen or printed), the analysis of the global autocorrelation (Moran I index) and of the correlogram for the first n lag, and the identification of local maxima and spatial clusters. Here the lag was calculated using the average diameter of the municipality, assuming that all municipalities were circular in shape, whilst the choice of the number of lags becomes a parameter of choice for the user. For a fuller discussion on the subject, we refer the reader to Cislaghi et al. (1995b), while for a general introduction to methods of kernel estimation and autocorrelation see Bailey and Gatrell (1995).

Particular attention must be given to two caveats. First, the atlas is based on the notifications of death on the death certificates, which may not relate to the underlying cause. Second, it is a purely descriptive instrument and as such has to be used for exploratory purposes rather than for confirmatory investigations. Furthermore, for a correct interpretation of the atlas outputs one must bear in mind the following points:

1. Each analysis of a circle is independent from the others as it has a different reference standard. This explains why one cannot compare the SMRs and the SPMRs of one circle with those of an other. To do so, all the data would have to be worked out again taking Italian population as the standard. But, in doing so, the analyses would not have been corrected for the spatial trend, nearly always present in the Italian geography of mortality.

2. A consequence of the above observation is that two identical colours in two different maps do not have the same values. Furthermore, the creation of the

groups represented in the graphs assumes a percentile division and as such relates to one map only.

3. The main aim of the atlas is to identify those areas with an anomalous concentration of cases and the analyses are carried out with this in mind. It is not useful therefore to know the geography of Italian mortality on a provincial or regional scale, for which there already exists many other descriptive studies.

4. The atlas is a geographical instrument and not an inquiry regarding individual municipalities. This means that more importance has been given to the study of the spatial structures rather than the values of single municipalities.

5. The kernel density estimates are based on a variable window size which takes into account the geographical sparseness of deaths.

6. The circle radius is set to a default value so that most of the times a circle includes a territory wider than the regional administrative borders. This choice was made to reduce the problem of biased estimates at the boundaries.

9.3 System design

The current version of the atlas – thanks to the parameterisation of the chosen variables and the control of this choice by a graphic interface (to be discussed in the following paragraph) – allows one to carry out the calculations on a personal computer. Therefore, while we retain the label 'atlas', we are dealing with a piece of interactive software, which can be easily used by experts in the field and which allows the user to carry out the calculations as they think best.

This system was produced thanks to the use of the product SAS and more precisely the modules SAS/BASE, SAS/GRAPH, SAS/IML, SAS/AF. The software was designed to run on a 486/66D PC with 16M of RAM, being careful to provide WINDOWS 3.1 with at least 15M of virtual memory. The actual version of the electronic atlas requires at least a 486 processor with a minimum of 8 Mb of RAM but, in order to work properly, a Pentium with 32 Mb of RAM is desirable.

SAS/BASE was used for the transformation of the raw data into DATA SET SAS and for the creation of index files for quick access to the information. SAS/IML was chosen for its management and calculating capacities on matrix/arrays so as to change the statistical calculations originally written in FORTRAN efficiently. SAS/GRAPH was used in the production of the mortality graphics (firstly written in characters and only then implemented in MAPINFO). SAS/AF, which is a component of the SAS system used to create user friendly windowing application, was utilised to produce – thanks to the use of the SCL language – a user interface. The final version of the product, including all the data files, will be released on CD-ROM. This is because of the amount of memory needed for the storage of the data (about 300M).

SAS was chosen as a development vehicle because of the following characteristics:

- The ease with which it can transform data into indexed charts (data set) and therefore being easily accessible through a SQL query.
- The possibility of parameterising all the data to be used.
- The existence of the programming language SAS/IML.

The quality of an application, in the field of computers, must today depend also on its visual appearance; that is, how it presents itself to the end user. The tools used in the project allowed the production of a valid user interface, facilitating and rendering both safe and immediate the use of the procedures which make up the application itself. The programming of the interactive functions between the end user and the SAS procedures of data processing was solved using the programming language SCL (Screen Control Language) SAS. With regards to the graphic part of the project, the SAS/AF software was used, which has SCL as the basis for its applications. From the software SAS/AF, the FRAME inputs were employed for the construction of graphic user interfaces (GUI) using the elements which are the most common in this kind of environment: bit map graphs, icons, pull down menus, menu buttons, scroll bars, sliders, radio boxes, check boxes and so on.

The user interface works at different levels which are present in the main menu where – in addition to being able to choose which SAS procedure you want to use (after having indicated the parameters requested) – you can also establish a 'Study location' (a directory on hard disk) in which the results of the calculation itself are stored as files.

9.4 Using the interactive atlas

Use of the system typically involves the following phases:

Main menu
Selection of the research area
Indication of the causes of death
Statistical data processing
Management of the result directories.

Before analysing in detail these steps, it is necessary to describe the solution adopted in the management of the files that are produced as the final result or as data files for future calculations.

The application is divided into levels which are expressed by a structure of directories, produced and managed on the user's hard disk. These directories are subsequently filled with the files produced from the single SAS procedures.

As a first step, the user singles out a 'Study location' which is stored in a main directory on the hard disk. Other sub-directories are then listed under such a directory, one for each chosen municipality. The calculation files concerning the research areas are then stored in this sub-directory. For each sub-directory of the centroids other sub-directories are created, one for every cause of death being studied. Finally, each sub-directory of causes contains further sub-directories, one for each statistic carried out on the data of the indicated cause.

The structure is stored in DATA SET SAS (SAS archive) which is listed under the 'Study location' directory. This data set produces selection lists which are proposed each time to the user and allow a choice to be made of data for each step in the session.

At the start of the application, there appears a menu which allows the following operations:

Definition of the 'work area'

Choice of the centroid municipality
Indication of the cause(s) of death
Statistical data processing
Management of the directories.

Before going on to any kind of processing, the user must indicate the study area which is going to be the subject of inquiry. Subsequently, from the main menu, one can activate the various research phases on the mortality atlas, phases which will now be described (while noting that they may undergo slight change as development work proceeds).

The first option in the main menu allows the user to define some preliminary parameters for any subsequent operation and processing. First, an indication of the centroid municipality of the study location under analysis is asked for. The choice is carried out by referring directly to the list of Italian municipalities. This operation allows rapid research and selection functions. This first phase finishes with the confirmation of the data here described. This operation is obligatory for the creation of the directory which will store the subsequently processed files. The second phase presents another series of values needed in the definition of the research area, connected to the previously selected centroid; in particular, the census to be referred to (currently 1981, but with an option for 1991) and the range determining the extension of the area itself. The range can be expressed in various forms: one can specify the maximum range of the circle around the centroid, the maximum number of municipalities to be analysed surrounding the centroid, the first municipality to be excluded in the selection, the municipality which represents the limit of the circle, or the choice of area around the centroid in relation to the number of inhabitants. With an Enter key, the processing of this first phase begins and data files concerning the research area under investigation are produced, on which the processing of the following stages will be based.

The second option in the main menu indicates – in a previously defined research area – the year to be considered and the cause of death. As such, the first piece of information to be provided by the user is the research area required out of those established in the previous step. This selection comes about by taking the data from the previously calculated list of research areas which automatically appears on the screen. The name to be given to the directory (causes directory) is asked for, the procedure stores the files connected with the processing and this is activated by an Enter key. It is useful to remember at this stage that one can process other causes by maintaining unchanged the research area relative to the chosen centroid, but varying when appropriate the other parameters, including the name of the directory to be given to the new cause.

The third and final phase consists in the production of the final results: the statistical processing which extracts data from the files produced in the two previous stages. First, a choice of the research area and cause of death, which have been previously processed, is confirmed. The second part proposes a substantial quantity of values and choices, with the purpose of guiding the statistical application in the direction wanted by the user and consequently obtaining the desired results. The final screen which summarises all the user choices is presented in Figure 9.2. At the moment, all the menu labels are written in Italian. Some of the screen options (such as Bayesian methods) are not implemented in the current version but the updated version of the atlas will cover all the listed alternatives.

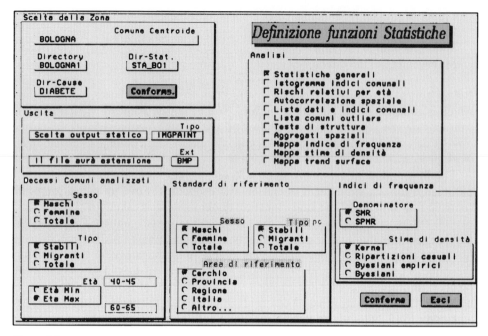

Figure 9.2

At this stage it is of no interest to study the single parameters which are required for the statistical processing. However, it may be useful to examine in more detail the compilation of the first section with regards to the structure of the directories. The user is asked to choose the previously processed research area, connected with a centroid. Subsequently, the user must choose the cause of death coupled with the chosen area and lastly insert the name to be given to the directory, in which will be stored the files concerning the final results for future statistical processing. These choices are guided by special selection lists which automatically appear on the screen.

Directory manangement, which can be activated from the main menu, is not related in any way to the processing procedures of the mortality atlas. Its aim is to administer the directories which are gradually built up and enlarged. When the user considers certain data to be no longer of any use, they can be deleted from the hard disk, so recovering precious space. The function allows the user to decide what is to be eliminated: a directory of research areas (centroid municipality), a directory of causes or statistics.

9.5 Case study

By way of two examples, we consider the frequency and density maps relating to the distribution of lung cancer mortality around one of the Tuscany Provinces (Lucca) and the geographical distribution of stomach cancer mortality in Lazio (centroid of Viterbo). The first example (Figure 9.3a) shows a very scattered spatial distribution of the Standardised Proportional Mortality Rates, with values ranging from 0 to

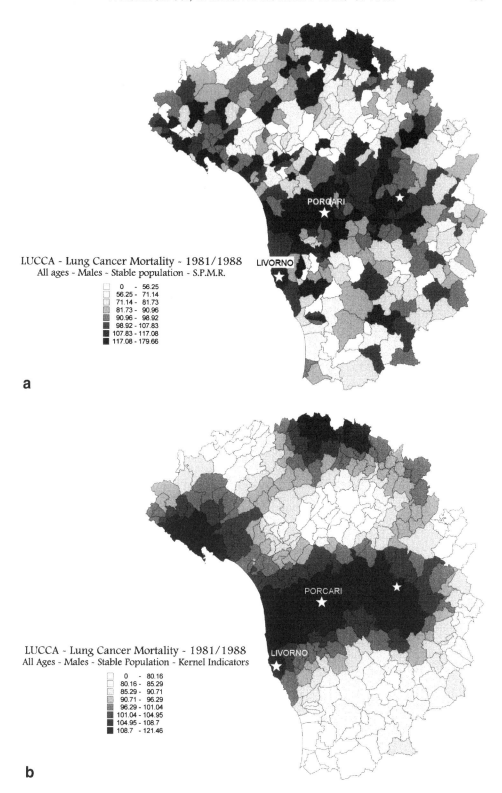

Figure 9.3 (a) Lucca – lung cancer mortality rates – 1981–1988, males, all ages, SPMR. (b) Lucca – lung cancer mortality rates – 1981–1988, males, all ages, kernel indicators.

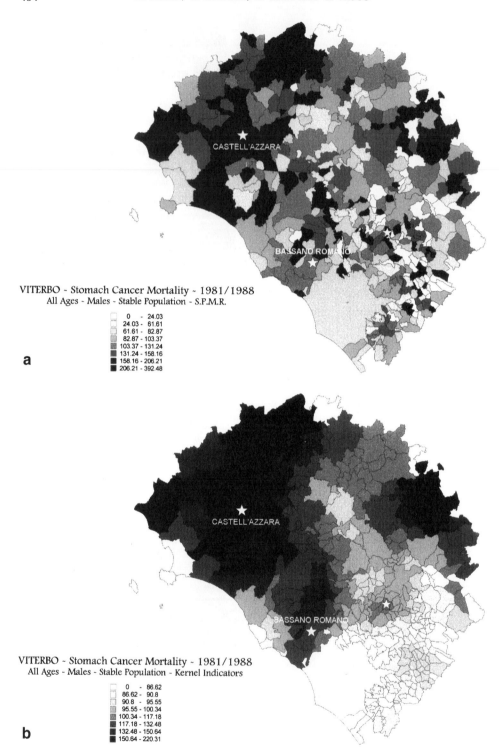

Figure 9.4 (a) Viterbo – stomach cancer mortality – 1981–1988, males, all ages, SPMR. (b) Viterbo – stomach cancer mortality – 1981–1988, males, all ages, kernel indicators.

179.7. On closer scrutiny it becomes evident that lung cancer mortality tends to cluster in the north-west and in the middle of the region. The north-west is the well known area of La Spezia which is a military harbour and where there is a concentration of chemical plants. The central area of the circle moving from the coastal area to the internal part of the region includes the main concentration of industrial activities in the Tuscany region. The most extreme east side of the high mortality area is centred on the city of Florence. The kernel map (Figure 9.3b) highlights in a clearer way these three areas, two of which have been identified as clusters, eliminating most of the small area variability. This gain in readability is balanced by the bias introduced by the smoothing procedure.

The second example (Figure 9.4a) shows the distribution of stomach cancer mortality in the Lazio Region. The high mortality is now located in the north-east part of the region and is connected with a well known area of high stomach cancer incidence which includes part of Tuscany and Emilia-Romagna. This cross-border area, affecting three different regions, has been the focus of many analytical studies which have considered the role of diet and genetic factors in the development of this cancer. Once again, the map displaying the smoothed kernel estimates (Figure 9.4b) provides a more readable image of the spatial distribution of the mortality for stomach cancer, at the expense of introducing some bias. How far the smoothed picture is from the real map depends on the correct choice of the spatial scale, the size of the kernel window and the strength of the underlying mortality structure compared with the background noise. However, the choice between real, although noisy information, and an artifactual but clearer picture is intrinsically linked with any statistical analysis which should be used with a deep knowledge of the methodology and results of which should be interpreted in the light of their scientific plausibility. In any case, the density estimation approach which is implemented in this atlas is, in our opinion, sufficiently easy to understand, flexible and based on few assumptions which are critical requirements for performing routine validation analyses on the consistency of the spatial patterns.

9.6 Conclusions

The need for a flexible and easy to use electronic atlas has been a matter of discussion for many years among the group of Italian researchers working on geographical epidemiology. The main concerns were the danger of misunderstanding the results of analyses based on aggregated data and the reduction in activity of public health services due to an excessive number of (possibly) false health alarms; in both cases the problem is due to a lack of specific training of the people responsible for the interpretation. A minor problem was the limited diffusion of computerisation in the peripheral health services. This latter aspect is no longer a problem since the reduction in prices of hardware and the increased power of the newest computers have overcome the resistance of the local bureaucracy. As for the problem of inexperienced people the solution was two-fold: to choose as default options those parameters offering low sensitivity to cluster identification and robustness of risk estimates; to train a group of people in each region which would have been responsible for using the electronic atlas and for interpreting the results of the analyses.

Another set of problems which were considered in the preliminary phase was which software to use. The idea of building a self-standing program was excluded from the beginning since it was considered too expensive. The choice of the SAS package was justified by the possibility of using it as a database and also as a global statistical package. The newest version of the electronic atlas and will use the GIS module of SAS. This will guarantee better quality of maps and the possibility of using the SAS facilities for screening the original data.

There is an ongoing evaluation of the use of the electronic atlas which consists in the follow-up of the people trained during the first phase of the project and the use of the atlas in four particular local health authorities of the Lombardy region (north of Italy). This latter is planned to evaluate the routine use of the atlas in small areas. The results of this evaluation process will be available shortly. The first version of the Italian Atlas is currently available on request only for users working in the Regional Health Planning institutions. However, once the problem of data confidentiality is solved, we will consider the possibility of extending the availability of the electronic atlas to other users.

References

BAILEY T. C. and GATRELL A. C. (1995) *Interactive Spatial Data Analysis*. Longman, Harlow.

BERNARDINELLI L. and MONTOMOLI C. (1992) Empirical Bayes versus fully Bayesian analysis of geographical variation in disease risk. *Statistics in Medicine*, **11**, 983–1007.

BITHELL J. F. (1990) An application of density estimation to geographical epidemiology. *Statistics in Medicine*, **9**, 691–701.

BRESLOW N. E. and CLAYTON D. G. (1993). Approximate inference in generalized linear mixed models. *Journal of the American Statistical Association*, **88**, 421–429.

CISLAGHI C., BIGGERI A., BRAGA M., LAGAZIO C. and MARCHI M. (1995a) Exploratory tools for disease mapping in geographical epidemiology. *Statistics in Medicine*, **14**, 2663–2682.

CISLAGHI C., BRAGA M. and BIGGERI A. (1995b). Analisi della concentrazione spaziale di eventi per mezzo delle superfici di densità. *Epidemiologia e Prevenzione*, **19**, 142–149.

GARDNER M. J., WINTER P. D., TAYLOR C. P. and ACHESON E. D. (1993) *Atlas of Cancer Mortality in England and Wales 1968–1978*. Wiley & Sons, Chichester.

JUNYAO L., BEQI L., HUANGYI L., SHOUDE R., et al. (1979) *Atlas of Cancer Mortality in the People's Republic of China*. China Map Press, Beijing.

KEMP I., BOYLE P., SMAMS M. and MUIR C. (1985) *Atlas of Cancer in Scotland, 1975–1980, Incidence and Epidemiologic Perspective*: IARC Scientific Publication No. 72, International Agency for Research in Cancer, Lyon.

MapInfo 2.1 – Desktop Mapping Software (1992). MapInfo Corporation, Troy, New York.

MARSHALL R. J. (1991) A review of methods for the statistical analysis of spatial pattern of disease. *Journal of the Royal Statistical Society A*, **154**, 421–441.

MASON T. J., MCKAY F. W., HOOVER R., BLOT W. J. and FRAUMENI I. F. (1975) *Atlas of Cancer Mortality for US Counties, 1950–1969*. US Deparment of Health, Education and Welfare, Bethesda.

SAS Institute Inc SAS/BASE, SAS/IML, SAS/GRAPH, SAS/AF. Version 10, Cary, NC, USA.

TASCO C., CISLAGHI C., BRAGA M. and BIGGERI A. (1993) Spatial components of variability in cancer mortality distribution. In *Statistics of Spatial Mortality Processes: Theory and Application*, 7–30 September, Bari.

VIGOTTI M. A. and CISLAGHI C. (1988). Cancer mortality in migrant populations within Italy. *Tumori*, **74**, 107–128.

WALLER L. A. and JACQUEZ G. M. (1995) Disease models implicit in statistical tests of disease clustering. *Epidemiology*, **6**, 584–590.

CHAPTER TEN

Bayesian Analysis of Emerging Neoplasms in Spain

GONZALO LÓPEZ-ABENTE

10.1 Introduction

Assessment of cancer mortality time trends in Spain has yielded evidence of a sharp increase in certain tumour sites. In the period 1982–1992, soft-tissue sarcomas, melanoma, non-Hodgkin's lymphoma (NHL) and multiple myeloma registered rises of over 70% in their respective age-adjusted rates for both sexes (López-Abente et al., 1995b). When mortality due to these causes was subjected to time-trend analysis employing age-period-cohort models, the trend was observed to contain a pronounced cohort effect, which led us to endeavour to monitor its evolution and pattern of spatial distribution (López-Abente et al., 1992; Pollán et al., 1993). As this phenomenon seems to affect other industrialised countries, there is now talk of emerging neoplasms (Carli et al., 1994; Hartge et al., 1994; Weisenburger, 1994), though it is still a moot point whether these are genuine increases in incidence or merely improvements in diagnosis and death certification.

Although these are relatively uncommon tumours, with an incidence in Spain of less than 5 cases per 100,000 inhabitants, were the exponentially increasing trend to continue unabated they would become a health problem of real concern. Table 10.1 shows the number of deaths and the adjusted mortality rates for 1992 (Centro Nacional de Epidemiolgía, 1995).

The risk factors for connective-tissue tumours, non-Hodgkin's lymphomas and multiple myeloma are unknown, yet by concentrating on the environmental components in these tumours, the involvement of chemical agents (pesticides, solvents, drugs) has been described, as has the increased risk in occupations linked to agriculture and cattle-farming (López-Abente, 1991; Demers et al., 1993; Storm, 1994).

In the study of the geographical patterns of these causes, low frequency in the study areas poses a problem even if Spanish provinces are used as the unit of analysis. Hence, the use of classical indicators (age-adjusted rates, standard mortality ratios (SMRs), rate ratios) may furnish unstable results, and the usefulness of smoothed estimators (empirical Bayes, full Bayes) in such situations has been

Table 10.1 Age-adjusted mortality rates (Spain 1992)

ICD	Cancer site	Cases	Age	ARE	ARW	Crude rate	% increase 1982–1992
Men							
162	Lung	13636	66.47	72.41	47.16	75.15	44
171	Connective tissue	194	55.15	1.17	0.99	1.07	70
172	Melanoma	294	59.64	1.59	1.11	1.62	156
200, 202	NHL	943	61.53	5.28	3.82	5.20	71
203	Myeloma	535	69.25	3.04	1.84	2.95	88
Women							
174	Breast	5617	63.75	27.57	18.60	30.67	36
171	Connective tissue	180	56.79	0.90	0.69	0.98	84
172	Melanoma	244	63.37	1.19	0.79	1.33	170
200, 202	NHL	760	66.63	3.58	2.34	4.15	102
203	Myeloma	535	70.53	2.45	1.49	2.92	87

ICD, International Classification of Diseases. ARE, age-adjusted rate (European population as standard). ARW, age-adjusted rate (World population as standard). NHL, non-Hodgkin's lymphomas.
Note: Rates for lung cancer in men and breast cancer in women are shown by way of quantitative reference.

acknowledged (Clayton and Kaldor, 1987; Bernardinelli and Montomoli, 1992; Cislaghi et al., 1995).

Similarly, an analytical approach has been formulated, based on the use of generalised linear mixed models (GLMM), which seeks to explain the geographical patterns and, in a highly efficient manner, to provide solutions to various problems plaguing classical analyses: spatial autocorrelated extra-Poisson variation, ecological regression analysis, and smoothed estimators for mapping. This chapter demonstrates an application of this method to connective-tissue tumours, multiple myeloma and non-Hodgkin's lymphomas, with two aims in mind:

1. To draw up smoothed maps for these diseases.
2. To seek variables capable of explaining their pattern of distribution.

10.2 Materials and methods

10.2.1 Mortality data

Based on individual records stored on magnetic tape and supplied by the National Statistics Office (Instituto Nacional de Estadística), deaths by age, sex and province were tabulated for the period 1988–1992 for the following rubrics of the *International Classification of Diseases* (ICD) (ninth edition): connective-tissue tumours (ICD 171), non-Hodgkin's lymphomas (ICD 200,202) and multiple myeloma (ICD 203). For the purposes of calculating expected cases (E_i) in each province, recourse was had to age-specific rates for Spain as a whole for the same period. Use was made of provincial populations by age group and sex as of 1 July 1990, calculated by interpolation of official censuses and municipal rolls.

10.2.2 Explanatory variables

The variables employed were the indicators for use of the following products (in tonnes): insecticides, acaricides (miticides), herbicides, industrial herbicides, molluscacides, and total pesticides (embracing the previous five) for 1989; fertilisers, and useful farm surface area relating to 1987. All the above indicators were categorised by quintile in order to employ an efficient method of categorisation capable of containing ten provinces in each interval. The source for the information on insecticides, acaricides, herbicides, industrial herbicides and molluscacides was the Business Association for the Protection of Plants (Asociación Empresarial para la Protección de las Plantas). Fertiliser data were taken from a report drawn up by the National Geographic Institute (Instituto Geográfico Nacional) on the basis of information furnished by the Ministry of Agriculture, Fisheries and Food.

10.2.3 Statistical analysis

To study the influence of the different covariates on the geographical pattern of mortality distribution, use was made of a model, initially proposed by Clayton and Kaldor (1987), and subsequently developed by Besag et al. (1991) and Clayton et al. (1993) in its application to ecological studies in epidemiology. This model is a Poisson GLMM (Breslow and Clayton, 1993) whose linear predictor contains a mixture of random and fixed effects. The model is expressed as:

$$\log(\theta_j) = \mu + \beta x_j + e_j^{[1]} + e_j^{[2]}$$

where the fixed-effects component (βx_j) corresponds to the explanatory variable(s) and the random-effects component breaks down into two parts: (1) variability, denominated heterogeneity ($e_j^{[1]}$), which is the unstructured extra-Poisson variation; and (2) influence of adjacency between areas, denominated clustering ($e_j^{[2]}$). The dependent variable is the area specific rate ratio (θ_j) controlled for age. The ratio of observed to expected cases (SMR = O_j/E_j) is the maximun likelihood estimate of θ_j.

The method for fitting the models adopts a Bayesian approach. A prior multivariate distribution is required for the model parameters to be specified. In line with previous proposals, uninformative hyperpriors for the random terms in the model were chosen (Bernardinelli et al., 1995). The prior probability distribution of the parameter values is converted into a posterior distribution of such values by using the data which are actually observed. Bayes theorem is used to derive the posterior distribution by means of combining the prior distribution with the likelihood of the data (Bailey and Gatrell, 1995). The model used represents each area as a sum of two random components that follow an unstructured and structured pattern of heterogeneity, respectively. The first component allows all relative risks to differ from one to another in an unstructured way. This assumes that relative risks are sampled independently from a prior distribution. The parameter is aspatial, in that it ignores the geographical location of any area. The second random component allows relative risks to vary in such a way that areas which are geographically close together will tend to have similar relative risks. A detailed formulation of the method will be found in the references (Besag et al., 1991; Breslow and Clayton, 1993; Clayton et al., 1993; Clayton, 1994; Bailey and Gatrell, 1995; Bernardinelli et al., 1995).

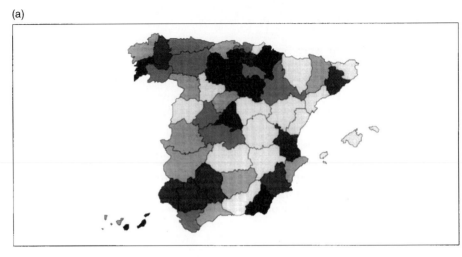
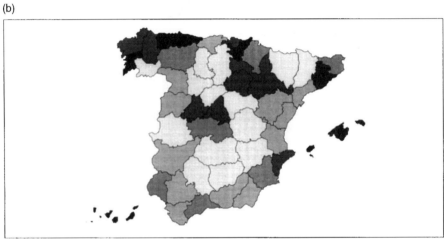
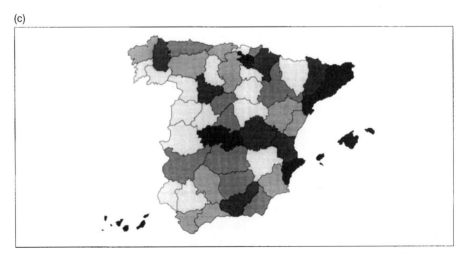

Figure 10.1 Provincial distribution of standard mortality ratios categorised by quintiles. (a) Connective tissue. (b) Non-Hodgkin's lymphoma. (c) Multiple myeloma. Men, Spain 1988–1992.

(a)
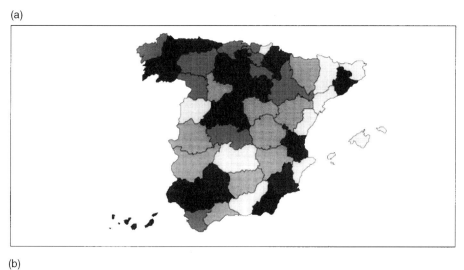

(b)
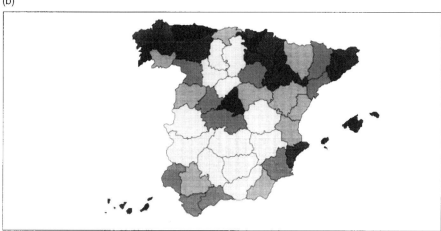

(c)
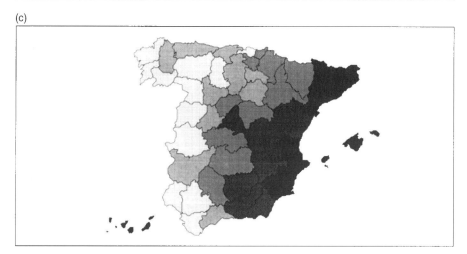

Figure 10.2 Provincial distribution of relative risks (full Bayes estimators) categorised by quintiles. (a) Connective tissue. (b) Non-Hodgkin's lymphoma. (c) Multiple myeloma. Men, Spain 1988–1992.

The Markov Chain Monte Carlo (MCMC) method as implemented in the BEAM (Bayesian Ecological Analysis and Mapping) programme (Clayton, 1994) was used for fitting the models, specifying a vague prior on the parameters. The programme option chosen was Gibbs sampling (Gilks and Wild, 1992). For the purposes of estimating the parameters, a minimum of 2000 samples were taken from all models after convergence had been reached. In addition, a Poisson model was used to ascertain changes in the estimation of the effect of the covariates and to illustrate the influence of overdispersion on such estimates.

10.3 Results

Shown in Figure 10.1 are the maps corresponding to SMRs calculated for the three causes studied, and in Figure 10.2, the smoothed estimators (full Bayes) without the inclusion of any explanatory variable. Figure 10.3 depicts the locations of the different provinces of Spain. A relative scale was used in these maps – the indicator being divided into quintiles – and hence the scale is different for each map. No defined pattern was observed for connective-tissue tumours, with Almería and Soria being the provinces registering the highest SMRs. The provincial range was 0.19 (Salamanca) to 1.69 (Soria). The use of the Bayes estimator in this cause brought about a drastic shrinkage in the indicator, grouping the relative risks (RR) around unity. The provincial range was 0.97 (Balearic Isles) to 1.08 (Las Palmas). The ensuing map was markedly similar to the first. The provinces with highest risk were the Canary Islands (RR 1.08 and 1.07) and Madrid (RR 1.05).

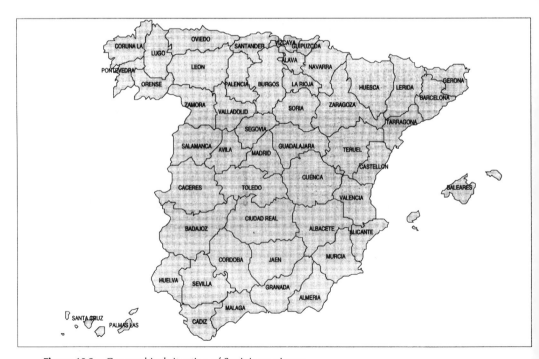

Figure 10.3 Geographical situation of Spain's provinces.

Table 10.2 Estimated posterior means (β) and posterior standard deviations (SD). Variables are categorised by quintile

Model	Connective tissue		Non-Hodgkin's lymphoma		Myeloma	
	β	SD(β)	β	SD(β)	β	SD(β)
Pest†	0.0156	0.0263	−0.0231	0.0119	0.0480	0.0158
Pest + Heter + Clust	0.0188	0.0289	−0.0175	0.0305	0.0319	0.0243
Pest + UFSA + Heter + Clust	0.0238	0.0306	−0.0115	0.0212	0.0404	0.0240
Insect†	0.0431	0.0239	−0.0229	0.0106	0.0424	0.0143
Insect + Heter + Clust	0.0514	0.0279	−0.0171	0.0284	0.0485	0.0205
Insect + UFSA + Heter + Clust	0.0631	0.0288	0.0008	0.0214	0.0629	0.0188
Acaric†	0.0215	0.0257	−0.0067	0.0116	0.0505	0.0154
Acaric + Heter + Clust	0.0211	0.0272	0.0254	0.0324	0.0328	0.0238
Acaric + UFSA + Heter + Clust	0.0304	0.0298	0.0154	0.0214	0.0464	0.0226
Herbic†	0.0140	0.0253	−0.0703	0.0114	0.0012	0.0151
Herbic + Heter + Clust	0.0190	0.0281	−0.0467	0.0232	0.0185	0.0211
Herbic + UFSA + Heter + Clust	0.0374	0.0317	−0.0157	0.0243	0.0320	0.0230
Hindus†	0.0231	0.0231	0.0223	0.0106	−0.0134	0.0137
Hindus + Heter + Clust	0.0269	0.0262	0.0251	0.0205	0.0040	0.0194
Hindus + UFSA + Heter + Clust	0.0271	0.0253	0.0300	0.0175	0.0034	0.0193
Molusc†	0.0418	0.0265	0.0768	0.0123	0.0437	0.0155
Molusc + Heter + Clust	0.0453	0.0283	0.0683	0.0208	0.0427	0.0187
Molusc + UFSA + Heter + Clust	0.0478	0.0330	0.0471	0.0214	0.0426	0.0223
Fertil†	0.0052	0.0248	−0.0730	0.0114	0.0047	0.0149
Fertil + Heter + Clust	0.0074	0.0265	−0.0653	0.0211	0.0076	0.0197
Fertil + UFSA + Heter + Clust	0.0387	0.0370	−0.0140	0.0273	0.0446	0.0274

Pest, pesticides. Insect, insecticides. Acaric, acaricides. Clust, clustering. Herbic, herbicides. Heter, heterogeneity. Hindus, industrial herbicides. Molusc, molluscacides. Fertil, fertilisers. UFSA, useful farm surface area.
† Maximum likelihood estimation. In that case, β is the regression coefficient and SD(β) the standard error.
In Bayesian analysis the posterior mean is the equivalent to the regression coefficient and the posterior SD to the standard error.

Table 10.3 Estimate of relative risk and its 95% credible interval (CI) for the variables studied categorised by quintile

Model	Connective tissue		Non-Hodgkin's lymphoma		Myeloma	
	RR	95% CI	RR	95% CI	RR	95% CI
Pest†	1.016	(0.965–1.069)	0.977	(0.955–1.000)	1.049	(1.017–1.082)*
Pest + Heter + Clust	1.019	(0.963–1.078)	0.983	(0.926–1.043)	1.032	(0.984–1.083)
Pest + UFSA + Heter + Clust	1.024	(0.964–1.087)	0.989	(0.948–1.031)	1.041	(0.993–1.091)
Insect†	1.044	(0.996–1.094)	0.977	(0.957–0.998)*	1.043	(1.014–1.073)*
Insect + Heter + Clust	1.053	(0.997–1.112)	0.983	(0.930–1.039)	1.050	(1.008–1.093)*
Insect + UFSA + Heter + Clust	1.065	(1.007–1.127)*	1.001	(0.960–1.044)	1.065	(1.026–1.105)*
Acaric†	1.022	(0.972–1.075)	0.993	(0.971–1.016)	1.052	(1.021–1.084)*
Acaric + Heter + Clust	1.021	(0.968–1.077)	1.026	(0.963–1.093)	1.033	(0.986–1.083)
Acaric + UFSA + Heter + Clust	1.031	(0.972–1.093)	1.016	(0.974–1.059)	1.047	(1.002–1.095)*
Herbic†	1.014	(0.965–1.066)	0.932	(0.912–0.953)	1.001	(0.972–1.031)
Herbic + Heter + Clust	1.019	(0.965–1.077)	0.954	(0.912–0.999)	1.019	(0.977–1.062)
Herbic + UFSA + Heter + Clust	1.038	(0.976–1.105)	0.984	(0.939–1.032)	1.033	(0.987–1.080)
Hindus†	1.023	(0.978–1.071)	1.023	(1.002–1.044)*	0.987	(0.961–1.014)
Hindus + Heter + Clust	1.027	(0.976–1.081)	1.025	(0.985–1.067)	1.004	(0.967–1.043)
Hindus + UFSA + Heter + Clust	1.027	(0.978–1.080)	1.030	(0.996–1.066)	1.003	(0.966–1.042)
Molusc†	1.043	(0.990–1.098)	1.080	(1.054–1.106)*	1.045	(1.013–1.087)*
Molusc + Heter + Clust	1.046	(0.990–1.106)	1.071	(1.028–1.115)*	1.044	(1.006–1.083)*
Molusc + UFSA + Heter + Clust	1.049	(0.983–1.119)	1.048	(1.005–1.093)*	1.044	(0.999–1.090)
Fertil†	1.005	(0.958–1.055)	0.930	(0.909–0.951)*	1.005	(0.976–1.034)
Fertil + Heter + Clust	1.007	(0.956–1.061)	0.937	(0.899–0.976)*	1.008	(0.969–1.057)
Fertil + UFSA + Heter + Clust	1.039	(0.967–1.118)	0.986	(0.935–1.040)	1.046	(0.991–1.103)

Pest, pesticides. Insect, insecticides. Acaric, acaricides. Clust, clustering. Herbic, herbicides. Heter, heterogeneity. Hindus, industrial herbicides. Molusc, molluscacides. Fertil, fertilisers. UFSA, useful farm surface area. * $p < 0.05$.
† Maximum likelihood estimation.

Table 10.4 Estimate of relative contribution of the heterogeneity and clustering components to overall variability of the relative risks. Posterior means [a] with posterior standard deviations [b] in brackets

	Model without covariates				Model with covariates			
	Heterogeneity	a:b	Clustering	a:b	Heterogeneity	a:b	Clustering	a:b
Connective tissue	0.0432[a]	3.15	0.0474	2.48	0.0434	2.80	0.0523	2.35
	(0.0137)[b]		(0.0191)		(0.0155)		(0.0223)	
NHL	0.0659	1.50	0.1925	5.80	0.0906	2.48	0.1115	4.22
	(0.0437)		(0.0332)		(0.0364)		(0.0264)	
Myeloma	0.0428	2.97	0.1306	4.80	0.0498	2.23	0.1228	4.48
	(0.0144)		(0.0272)		(0.0223)		(0.0274)	

In non-Hodgkin's lymphomas, the SMR ranged from 0.41 (Palencia) to 1.87 (Las Palmas). There may be an inland-coast geographical pattern. On the smoothed maps, Palencia (RR 0.64) registered the lowest RR and Las Palmas (RR 1.82), the highest. Highest-risk provinces were Las Palmas, Santa Cruz de Tenerife, Barcelona, Pontevedra and Guipúzcoa. In the case of multiple myeloma, Zamora exhibited the lowest SMR (0.46) and Santa Cruz de Tenerife, the highest (SMR 1.61). The smoothed map showed Santa Cruz de Tenerife (RR 1.54) and Pontevedra (RR 0.83) as being the provinces with highest and lowest risk, respectively. No defined geographical pattern was evident on the unsmoothed map. On the smoothed map (Figure 10.2c), however, a clear west–east pattern was observable, with the Mediterranean coastal provinces (Catalonia, Balearic Isles, Valencian Region and Murcia) registering a higher mortality. Smoothing hardly affected the Canary Islands due to the fact that, insofar as the coding of adjacencies is concerned, the archipelago possesses no contiguous provinces.

In the ecological analysis presented here, an effort was made to find variables which might explain the distribution of the provincial RR. Table 10.2 sets out the effect estimators for the explanatory variables in three situations:

1. Fitting a classical log-linear Poisson model (Model 1).
2. Using a model which includes the two extra-Poisson components (Model 2).
3. Adjusting the estimator for useful farm surface area (Model 3).

Table 10.3 shows the same information in the form of relative risk and its 95% confidence intervals. The effect estimator indicates the change in risk by quintile of exposure (trend test). For connective-tissue tumours, the only variable evincing statistical significance was insecticide use in Model 3. Molluscacides showed association with mortality due to non-Hodgkin's lymphomas and multiple myeloma. Insecticides and acaricides appeared to be linked to multiple myeloma. The negative association between fertiliser and herbicide use and non-Hodgkin's lymphomas disappeared on adjusting for useful farm surface area.

Furthermore, Tables 10.2 and 10.3 illustrate the phenomena of extra-Poisson dispersion and confounding by location. There were no overdispersion phenomena in the Poisson model for connective-tissue tumours. Indeed, in this cause, the standard errors of the estimators in the two models (Poisson vs mixed model) for all variables are very similar (deviance 45 with 49 df). In NHL and myeloma, however, adjustment for extra-Poisson dispersion proved important, since statistical significance disappeared in the case of the insecticides and industrial herbicides variable. Modelling of variation by heterogeneity and clustering amounts, in some way, to including unmeasured explanatory variables in the model. The clustering component implies modelling the effect of geographical location. If the geographical variation of the variable studied is similar to that of risk of dying, location may act as a confounder (Clayton et al., 1993).

Table 10.4 shows the characteristics of the random heterogeneity and clustering components in models with and without explanatory variables. Both components contributed significantly to overall variability in provincial RR. The clustering component was greater than that of heterogeneity in NHL and myeloma. The interpretation put on the presence of this component is that it is evidence of possible unmeasured risk factors, which are a combination of environmental exposures and local socio-demographic features.

10.4 Discussion

The results point to the possibility that large-scale use of phytosanitary products (such as, insecticides, acaricides and molluscacides) in agriculture may have a measure of influence on the mortality of the tumours studied.

The validity and accuracy of Spanish mortality statistics, drawn up on the basis of death certificate data, have been studied on a number of occasions and their utility accepted, especially in the case of malignant tumours, with the information being of very high quality in tumours of the lymphatic and hæmatopoietic systems (Benavides et al., 1989). Part of the observed increase in mortality rates may be attributable to diagnostic improvements, yet the most widely-held opinion is that such improvements only explain a small proportion of the said rise (Pollán et al., 1993; Weisenburger, 1994), though the situation may conceivably be different in the three tumours in question. With respect to geographical distribution, there are no reasons to suppose that, in the five years covered by this study, marked health care differences might have been present in the various Spanish provinces which would have accounted for the patterns – and different patterns at that – observed for the three causes studied.

The use of SMRs as an indicator of mortality entails the difficulty of their not being strictly comparable, owing to the use of different standard populations. Despite this, the pattern obtained when these are plotted on maps resembles that obtained with the adjusted rates by the direct method (López-Abente et al., 1995a). In Spain, the availability of geographically-based indicators for agricultural chemicals used is very limited. The data used for this study were furnished by the Spanish Agrochemical Producers' Association, the only body which has records specifying product type and use at a provincial level. These data are compiled as a summary of sales and, though it is not possible to assess the quality of same, there is no reason to doubt their reliability. The source of data does not allow for the chemical composition of the various products to be known.

Ecological studies are nonetheless a useful exploratory tool, allowing for analysis, on a very preliminary basis, of environmental health problems difficult to tackle with other designs (Clayton et al., 1993). Indicators of chemical substance use for areas of any size below the provincial level were not available to us. Use of the smallest possible spatial unit would imply a greater degree of intra-area homogeneity and a reduced possibility of ecological bias (Greenland, 1992). Prolonged latency periods were not taken into account in the analysis; the indicators referred to 1987 and 1989, and the median year for deaths was 1990. Accordingly, work proceeded on the assumption that the information for the 1987–1989 period was an indicator of the use of these substances in preceding decades and that inter-provincial differences had remained steady over time, an assumption that is in no way rash, bearing in mind that crop types and treatment traditionally tend to remain unchanged for years at a stretch. This implies the presence of inevitable errors of classification in most ecological studies.

The mapping of geographical variation of risk for a given disease based on mortality rates in small geographical areas is very important in the production of disease atlases and ecological studies aimed at advancing aetiological hypotheses (Clayton and Kaldor, 1987; Bernardinelli and Montomoli, 1992; Clayton et al., 1993; Bernardinelli et al., 1995; Cislaghi et al., 1995). However, when the disease is rare (or the spatial unit small), direct representation of data (rates) may reflect an unreal image of their heterogeneity, proving very difficult to interpret. A typical characteristic of such maps is that their most extreme values are very imprecise. An approximation by means of Bayesian modelling enables any variation in the map to be 'filtered' and a picture obtained which better reflects the true heterogeneity of the risk (Clayton and Kaldor, 1987; Bernardinelli and Montomoli, 1992; Calyton et al., 1993; Bernardinelli et al., 1995; Cislaghi et al., 1995). The model used, based on Clayton and Kaldor's (1987) initial proposal was adapted by Besag et al. (1991), introducing concepts developed in the area of image analysis (each area can be viewed as a pixel). There are a number of advantages in applying smoothing techniques in these situations. The model used includes two random-effects terms ($e_j^{[1]} + e_j^{[2]}$) which can be interpreted as surrogates of unknown and unobserved variables. As mentioned above, the term $e_j^{[2]}$ would represent variables with a certain spatial structure, for example variables which, insofar as they correspond to contiguous areas, would be more likely to bear a resemblance than would other areas that are far apart and are in no way related. The inclusion of $e_j^{[1]}$ is ascribable to Breslow who furnished evidence of extra-Poisson variation in basic log-linear models (Breslow, 1984). Breslow and Clayton bracketed this model in a class which they denominated generalised linear mixed models (Breslow and Clayton, 1993). Through inclusion of random-effects terms, the model employed in this analysis combines the benefits of correcting for extra-Poisson dispersion, spatial autocorrelation and confounding by location (Clayton et al., 1993). The inclusion of the clustering term allows the effect estimate to be adjusted by location, thus to some extent controlling for possible confounding effects of unquantified variables (Clayton et al., 1993). Such an adjustment would be equivalent to matching by place of residence in other designs. In the tumours under review, inclusion of the heterogeneity and clustering terms fails perceptibly to modify the RR for the exposures studied. Correcting for extra-Poisson dispersion has clearer consequences on statistical significance in insecticides and industrial herbicides in the case of NHL, and the pesticides variable in that of myeloma.

For a long time, use of full Bayes estimators has not, in practice, been feasible because of computational intractability. Recent advances in MCMC methods for Bayesian estimation enable the application of these methods to different problems. There are basically two MCMC interactive simulation methods, that is Gibbs and Metropolis-Hasting samplers. A fuller description of the method and its possible uses will be found in Spiegelhalter et al. (1995) and Gilks et al. (1995).

The results reveal an association between: insecticide use and connective-tissue tumours and myeloma; molluscacides and non-Hodgkin's lymphomas; and, acaricides and myeloma. Herbicide use does not appear to be associated with any of the causes studied, though its association with soft-tissue sarcomas has been described (Hoar et al., 1986). We failed to find references in the literature to the carcinogenicity of molluscacides and acaricides. The results relating to the pesticide variable suggest that, when studying the effects of these substances, it is important for the nature of exposure to be specified.

The range of the effect estimators is considerably reduced by the use of full Bayes estimators, which smooth any variation. The use of a fixed scale in graphic representation hinders the possibility of appreciating a gradient pattern in the maps, which is why it was decided to represent distribution in quintiles. While smoothing of the maps by means of full Bayes estimators did not render the presence of patterns in connective-tissue tumours and NHL any clearer, in myeloma the process nevertheless revealed a manifest west–east gradient, with highest risk being registered for the Mediterranean coastal provinces. This pattern coincides visually with that plotted for the use of molluscacides, herbicides, acaricides and insecticides. The high mortality in the Canary Island provinces is remarkable. Of those studied, the only exposure identified in these provinces was large-scale use of acaricides.

The methodology and results of this study are of a preliminary, exploratory nature. The MCMC method allows for great flexibility in the formulation of models, and software is currently being developed which will facilitate use of same. The following would prove useful: automation of convergence criteria and number of samples required; generalisation of the model, thereby enabling all types of variables to be worked with; automatic creation of dummy-variables; and the study of interactions (Spiegelhalter *et al.*, 1995). We feel that it would be of great interest if these analytical techniques could be developed, applied to real problems in epidemiology, disseminated throughout the sector conducting research into spatial analysis of health problems, and included in GIS packages so as to enable use of same for routine analysis in spatial disease distribution studies.

Acknowledgments

Grateful thanks must go to Marina Pollán for her advice and help, and to María Ruiz Tovar for her comments and for furnishing definitive data on the indicators studied.

References

BAILEY T. C. and GATRELL A. C (1995) *Interactive Spatial Data Analysis*. Longman, Harlow.

BENAVIDES F. G., BOLUMAR F. and PERIS R. (1989) Quality of death certificates in Valencia, Spain. *American Journal of Public Health*, **79**, 1352–1354.

BERNARDINELLI L., CLAYTON D. and MONTOMOLI C. (1995) Bayesian estimates of disease maps: How important are priors? *Statistics in Medicine*, **14**, 2411–2431.

BERNARDINELLI L. and MONTOMOLI C. (1992). Empirical Bayes versus fully Bayesian analysis of geographical variation in disease risk. *Statistics in Medicine*, **11**, 983–1007.

BESAG J., YORK J. and MOLLIÈ A. (1991) Bayesian image restoration, with applications in spatial statistics (with discussion). *Annals of the Institute of Statistics and Mathematics*, **43**, 1–59.

BRESLOW N. E. (1984) Extra-Poisson variation in log-linear models. *Applied Statististics*, **33**, 38–44.

BRESLOW N. E. and CLAYTON D. G. (1993) Approximate inference in generalized linear mixed models. *Journal of American Statistical Association*, **88**, 9–25.

CARLI P. M., BOUTRON M. C., MAYNADIE M., BAILLY F., CAILLOT D. and PETRELLA T. (1994) Increase in the incidence of non-Hodgkin's lymphomas: evidence for a recent sharp increase in France independent of AIDS. *British Journal of Cancer*, **70**, 713–715.

Centro Nacional de Epidemiología. Mortalidad por cáncer. España 1992. (1995) (http://193.146.50.130/cancer/mort92.txt).

CISLAGHI C., BIGGERI A. and BRAGA M. (1995) Exploratory tools for disease mapping in geographical epidemiology. *Statistics in Medicine*, **14**, 2363–2381.

CLAYTON D. (1994) *BEAM: A program for Bayesian Ecological Analysis and Mapping*. Version 2.02, MRC Biostatistics Unit, Cambridge.

CLAYTON D., BERNARDINELLI L. and MONTOMOLI C. (1993) Spatial correlation in ecological analysis. *International Journal of Epidemiology*, **22**, 1193–1202.
CLAYTON D. and KALDOR J. (1987) Empirical Bayes estimates of age-standardized relative risks for use in disease mapping. *Biometrics*, **43**, 671–681.
DEMERS P. A., VAUGHAN T. L., KOEPSELL T. D., LYON J. L., SWANSON G. M., GREENBERG R. S. and WEISS N. S. (1993) A case-control study of multiple myeloma and occupation. *American Journal of Industrial Medicine*, **23**, 629–639.
GILKS W. R., RICHARDSON S. and SPIEGELHALTER D. J. (1995) *Markov Chain Monte Carlo in Practice*. Chapman and Hall, London.
GILKS W. R. and WILD P. (1992) Adaptive rejection sampling for Gibbs sampling. *Applied Statistics*, **41**, 337–348.
GREENLAND S. (1992) Divergent biases in ecologic and individual-level studies. *Statistics in Medicine*, **11**, 1209–1223.
HARTGE P., DEVESA S. S. and FRAUMENI J. F. JR (1994) Hodgkin's and non-Hodgkin's lymphomas. *Cancer Survey*, **19–20**, 423–453.
HOAR S. K., BLAIR A., HOLMES F., BOYSEN C. D., ROBEL R. J., HOOVER R. and FRAUMENI J. F. JR (1986) Agricultural herbicide use and risk of lymphoma and soft-tissue sarcoma. *Journal of the American Medical Association*, **256**, 1141–1147.
LÓPEZ-ABENTE G. (1991) *Cáncer en Agricultores. Mortalidad Proporcional y Estudios Caso-Control con Certificados de Defunción*. Fondo de Investigación Sanitaria, Madrid.
LÓPEZ-ABENTE G., POLLÁN M., ESCOLAR A., ERREZOLA M. and ABRAIRA V. (1995a) *Atlas de mortalidad por cáncer y otras causas en España* (http://www.uca.es/atlas/introdu.html).
LÓPEZ-ABENTE G., POLLÁN M. and RUIZ M. (1995b) El cáncer, un problema de salud prioritario. *Boletín Epidemiológico Semanal*, **3**, 81–84.
LÓPEZ-ABENTE G., POLLÁN M., RUIZ M., JIMÉNEZ M. and VÁZQUEZ F. (1992) *Cancer mortality in Spain, 1952–1986. Effect of age, birth cohort and period of death*. Centro Nacional de Epidemiología, Madrid.
POLLÁN M., LÓPEZ-ABENTE G. and PLÁ R. (1993) Time trends in mortality for multiple myeloma in Spain, 1957–1986. *International Journal of Epidemiology*, **22**, 45–50.
SPIEGELHALTER D., THOMAS A., BEST N. and GILKS W. (1995) *BUGS: Bayesian inference using Gibbs sampling*. Version 0.50, MRC Biostatistics Unit, Cambridge, (ftp://ftp.mrc-bsu.cam.ac.uk).
STORM H. H. (1994) Cancers of the soft tissues. *Cancer Survey*, **19–20**, 197–217.
WEISENBURGER D. D. (1994) Epidemiology of non-Hodgkin's lymphoma: recent findings regarding an emerging epidemic. *Annals of Oncology*, **5** Suppl. 1, S19–S24.

CHAPTER ELEVEN

The Development of an Epidemiological Spatial Information System in the Region of Western Pomerania, Germany

NANJA VAN DEN BERG

11.1 Introduction

This chapter describes the conception and initial use of a GIS for epidemiological research, developed as a cooperative project between the Spatial Data Centre (ZRI) and the Institute of Hygiene and Environmental Medicine at Greifswald University. The project, which commenced in March 1996, aims to develop an epidemiological spatial information system, useable in the first instance by epidemiologists in the Institute of Hygienic and Environmental Medicine.

The project, known as ERIS (an Epidemiological Spatial Information System) consists of three stages of development. First, the visualisation of epidemiological data, database query, and the development of a user interface (March 1996–January 1997). Second, the development of spatial analysis and geostatistical tools (February 1997–January 1998). Third, the integration of a dynamic element (probably February 1998–January 1999). The first two stages are discussed below. There is, as yet, relatively little attention devoted to incorporating a time element into epidemiological GIS (though see Löytönen, chapter 7, for some pointers to further research). The present chapter reviews some methods and problems encountered in the early stages of the ERIS project.

The research region of Western Pomerania is situated in the federal state of Mecklenburg-Vorpommern, in the former GDR. It is, in German terms, a poor region with an unemployment rate of more than 20% and inadequately developed infrastructure. Western Pomerania has two cities of approximately 65 000 inhabitants (Stralsund and Greifswald), but only Greifswald is in the research region. Otherwise, settlement is dispersed; the region is primarily agricultural, with little important industry.

11.2 The epidemiological database

Western Pomerania participates in the world-wide ISAAC study (ISAAC = International Study of Asthma and Allergies in Childhood), which has involved data collection in 48 countries (Asher et al., 1995). The goals of ISAAC are:

1. The collection of data about frequency and seriousness of asthma and allergies in children of 6–7 and 13–14 years old under different living conditions and in different countries.
2. The collection of basic epidemiological data, in order to make predictions about variations in frequency and seriousness of these illnesses in future years.
3. The development of a worldwide framework for future aetiologic research, examining links with genetics, lifestyle, environmental factors and medical care.

The data available for ERIS exist in the form of records for about 6000 children: 3000 children between the ages of 13 and 14 years filled in a questionnaire, while 3000 parents did the same for their children aged between 7 and 8 years. The children and parents were mainly asked for skin complaints and breathing difficulties, the condition and location of their homes, an estimate of the number of cars driving past their homes, and the presence of waste gases in the indoor environment. The symptoms under investigation were described in the questionnaires and in some cases were shown on video.

The age group of 13–14 year-old children was chosen to reflect the fact that mortality from asthma can be a problem; the use of both a self-completed questionnaire and a video questionnaire is possible for this group. The age group of 6–7 year-old children was chosen to give a reflection of the early childhood years, when asthma tends to be most prevalent, and when hospital admission rates are higher (Asher et al., 1995).

The ISAAC study has attracted some critical attention since all answers are subjective: no doctors, epidemiologists or environmental specialists were involved in making any diagnosis. In this inquiry only the perceptions of the children (and possibly parents) were important. It is difficult to judge what influence these forms of data acquisition have on the data reliability; are children (and parents) able to relate the described symptoms to their own experience? The same problem arises from the questions about environmental factors; although it is interesting to find out how children perceive their environment, it is impossible to standardise these perceptions. Fortunately, it will shortly be possible to compare the experiences of the childrens' environmental circumstances with real data, collected during other research projects.

The dataset is quite large, but does not cover the whole area of Western Pomerania. The children were recorded at every school that gave permission. Most of the schools agreed to the study, but for unknown reasons none of the schools in the two large towns of Bergen and Sassnitz were questioned and one of the schools in the city of Greifswald refused permission. The method of questioning children at the schools has the advantage that a lot of children can be reached quite easily and fast, but has also two disadvantages. First, certain areas are represented disproportionately, which could have an impact on the occurrence of allergies. Second, there are potentially problems in border areas, since children living in such areas could go to school outside the research region.

The response rate to the questionnaires was approximately 86%. From these questionnaires, about 5% were not useable for various reasons, such as incompleteness or non-existent addresses. Despite these shortcomings, for the region of Western Pomerania as a whole the dataset is representative and gives a good picture of the incidence of asthma and allergies in children. However, splitting the data into smaller areas, for example ZIP-codes, is problematic because of small numbers. An exception is the city of Greifswald, where, in each age group, approximately 900 children were questioned.

To date, six variables (the same in both age groups) have been chosen for the visualisation and spatial analysis. These variables correspond to the following questions: whether asthma symptoms were reported during the last 12 months; whether a doctor diagnosed asthma; whether hay fever symptoms were reported during the last 12 months; whether a doctor diagnosed hay fever; whether symptoms of endogenous eczema during the last 12 months were reported; and whether a doctor had diagnosed endogenous eczema.

11.3 Visualisation of epidemiological data

Visualisation of the data is usually the first step in both epidemiological and other spatial investigations, though there will be a close interaction between this and exploratory data analysis and modelling (Gatrell and Bailey, 1996). A visualisation of the raw or only slightly adjusted data can be very useful in obtaining a first overview of the spread of data and possible spatial patterns.

An important question is which geographical scale is most appropriate in representing epidemiological data. Of course, this depends on the nature, density and geographical referencing of the data. An example from the cancer registry in the former GDR (Möhner and Stabenow, 1994) shows how important the choice of the right geographical unit is. In the area of the former GDR there were about 7500 communities. Some 85% of them had less than 2000 inhabitants. In these rural villages lived about 23% of the total population. With approximately 60 000 new cancer illnesses every year, that meant an average of two cases in each rural village. This suggests that the communal level is not especially useful for visualising cancer rates in the former GDR.

Often, maps are made for people without a great knowledge of cartography, geography or GIS; and this group will often include epidemiologists and public health doctors! But these are the people who have to interpret such maps and use them to inform their decision making, for example about prevention strategies. This means that collaboration between GIS specialists and epidemiologists is absolutely essential.

As different authors have already noticed, the visualisation could be the most dangerous part of the whole procedure: scales, images and colours are not neutral and are easy to manipulate. If we understand the nature and perception of image and colour, we can use this to reinforce the selected message (Westlake, 1995). The interpretation and understanding of maps is both a perceptual and a cognitive process. There are characteristic ways in which humans will interpret a visual representation, some of which can lead to misinterpretation, and which need to be recognised (Hearnshaw, 1994). In this framework, the ideas of Edwards (1987, in Wood,

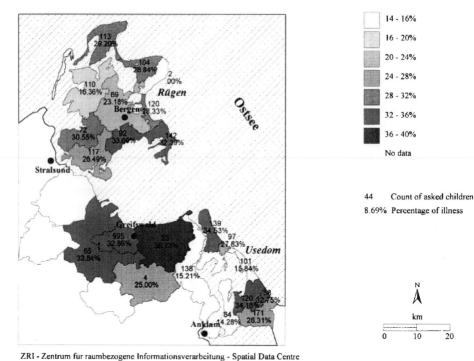

Figure 11.1 Hay fever symptoms during the last 12 months, by 12–14 year-old children.

1994) are very interesting to get an idea about the process of creative thinking. This process involves the following stages.

1. First insight: this is the exploratory stage, involving searching out productive questions from intuitive leaps of awareness based on wide knowledge of a subject. It can relate to existing problems or involve problem-finding.
2. Saturation: this forms the first stage of active research, involving gathering, sorting and categorising information with a particular end in view.
3. Incubation: when lines of investigation come to an end, logical analysis fails and frustration often results. The problem under consideration may seem to be set aside.
4. Illumination: this is where, normally quite suddenly, a solution may surface: the 'Ah-ha!' experience.
5. Verification: although the previous stage may be interpreted as just good luck, the outcome is accepted and the investigator proceeds to test the solution(s) against known information.

Of course, GIS specialists are not psychologists. The ideas above are only mentioned to indicate that map interpretation is a complex process which can easily be manipulated by using the right or wrong colours, symbols, scale, class intervals, and so on.

Turning to the ISAAC data, there are a few special problems, both of visualisation and subsequent spatial analysis, related to the spatial scale of the data. There are, in fact, two scales of analysis; the first level is the ZIP-code level, the second level is the address level for the city of Greifswald only. The German ZIP-code system consists of five-digit numbers. The last three numbers indicate the postal district. Normally, villages have one postal district, cities have two or more postal districts and small villages are aggregated to one postal district. The borders of the ZIP-code regions are fundamentally the same as the municipality borders.

The difference between these two levels is more than one of visualisation: research at the ZIP-code level will suggest different causes for allergies than research at the address level. At the ZIP-code level the observer might hypothesise, for example, that environmental factors such as chemical installations or the influence of the wind direction on an exposed area will be important. Looking at the address level in a city, one expects factors like the volume of traffic in a street or the city's micro-climate to be important. In summary, one can say that studies of health effects require environmental data on the same scale as that observed for the health data (Stern, 1995). In the ERIS investigation, the epidemiologists are searching for explanations at both levels.

The visualisation problems are most important at the ZIP-code level. Although the total number of children asked is quite large, the numbers in some ZIP-code regions are very small. That means that a responsible cartographic representation is only possible by omitting these areas, or by displaying the number of respondents (see Figures 11.1–11.4).

The second level is address-based: all children living in the city of Greifswald are recorded by unique xy-coordinates. Given such data, the first logical expression is a

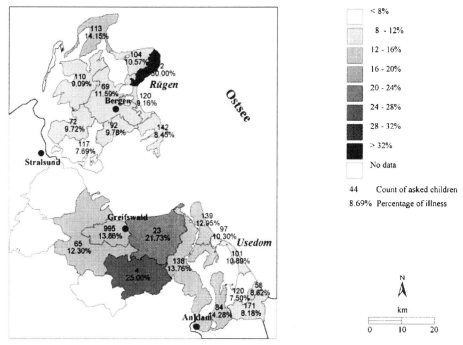

Figure 11.2 Diagnosis of hay fever, by 12–14 year-old children.

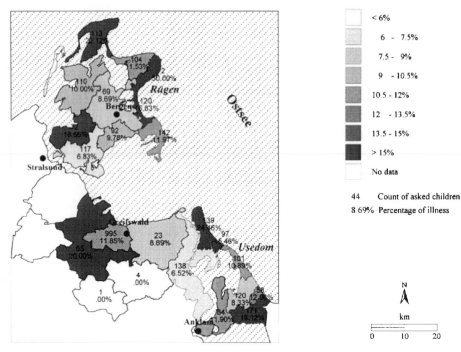

Figure 11.3 Asthma symptoms during the last 12 months.

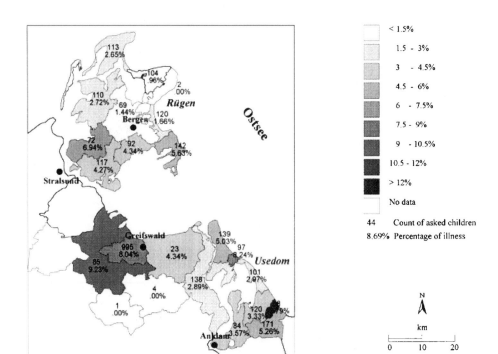

Figure 11.4 Diagnosis of endogenous eczema, by 12–14 year-old children.

dot map, which can give an impression of variations in the density of the objects in the study region and, since density is a continuous variable, it is also a starting point for an eventual transformation of the dot map into a continuous density surface (as shown, for example, by Gatrell, 1994). In this research project, visualisation and analysis of the data on an address-based level in Greifswald is useful, because the number of respondents is large enough (about 1000 in both age groups) and potential important environmental data, such as traffic counts, are also available at this scale.

11.4 Spatial analysis of the ISAAC data for Western Pomerania

11.4.1 Spatial analysis at the ZIP-code level

Like the visualisation, the spatial analysis of the data takes place at both ZIP-code and address-based levels. The first step is to find any spatial disease patterns. This leads to at least three problems. First, at the ZIP-code level we do not have the exact geographic locations of the children, instead we use the centres of the ZIP-code areas. Jacquez and Waller (1996) show in a HIV-simulation model that there is a substantial decrease in statistical power by increasing aggregation levels, and one of their conclusions is that actual clusters may be missed because of the loss of statistical power due to location uncertainty. Secondly, the ZIP-code regions are, as mentioned above, represented disproportionately: the number of children responding varies from just one in some rural border regions to almost 1000 in each age group in the city of Greifswald. Thirdly, the number of ZIP-code regions is very small: dependent on the chosen variable there are between 20 and 22 zones. The city of Greifswald has three ZIP-code areas, but because of the homogeneous character of the area and the fact that the borders of the original ZIP-code areas are quite arbitrary, the three areas have been aggregated to one region. Clearly, we need to bear these problems in mind and to interpret results of spatial analysis very carefully.

Useful under these circumstances is a chi-square analysis of the difference between observed numbers and the Poisson expectation (Blalock, 1981; Bortz, 1993; Brown et al., 1990), since all variables are rather rare events. In the 6–8 year-old age group, only hay fever and asthma as diagnosis are not Poisson distributed (Table 11.1). Nevertheless, it is not possible to make any statement with this result, since

Table 11.1 Age group 6–8 years

Variable	p	Degrees of freedom	χ^2	H_0
Diagnosis hay fever	0.028	3	14.15	rejected
Hay fever symptoms	0.094	14	23.05	not rejected
Diagnosis asthma	0.028	3	12.70	rejected
Asthma symptoms	0.067	11	11.88	not rejected
Diagnosis endogenous eczema	0.088	14	24.58	not rejected
Endogenous eczema symptoms	0.071	11	5.32	not rejected

H_0: the number of children (6–8 years) with these illness or symptoms is Poisson distributed over the ZIP-code areas.

there are only 5 ZIP-code areas with an expected count of cases over 5. Much better are the results in the 12–14 age group, where 4 of the 6 variables are not Poisson distributed (Table 11.2); these are: diagnosis hay fever; hay fever symptoms during the last 12 months; asthma symptoms during the last 12 months; diagnosis endogenous eczema.

In this case the statement can be made that the distribution of these 4 variables is not non-random. If we assume that the genetic conditions of all children are spatially independent, it is probable that large-scale environmental factors are responsible for the spatial pattern. Yet, it is interesting to take a look at the maps (Figures 11.1 and 11.2). The regions with the highest and lowest hay fever percentages (diagnosis and symptoms during the last 12 months) were not taken into account in the analysis since the expected count of cases was too small here. The statistically interesting regions are those of the category 12–16% by hay fever diagnosis and 32–40% by the symptoms. Of these regions, the city of Greifswald has by far the largest deviation from the expected count of cases. This suggests that there is a difference between the urban and the rural areas for the occurrence of hay fever in this age group.

The variable 'diagnosis endogenous eczema' also shows a difference between urban and rural areas: the city of Greifswald has the highest χ^2 and deviation from the expected count. On the map (Figure 11.4) the area around Greifswald shows an even higher percentage of cases, but this region was not taken into account, because the expected count of cases was under 5.

The variable 'asthma symptoms during the last 12 months' (Figure 11.3) points out a different pattern: here the city of Greifswald has a very small χ^2; no significant deviation from the expectation. The areas with the highest deviations from the expected counts are the regions in the north of the islands of Rügen and Usedom. These are also the most exposed areas of Western Pomerania. A climatic influence cannot be excluded here.

This first statistical analysis gives a few interesting and surprising results. A bit disappointing is the absence of any result in the age group with the 6–8 year-old children. Among 12–14 year olds, there is, as expected, in the urban part of the research area (the city of Greifswald), a higher potential of some allergy symptoms, namely hay fever and endogenous eczema. Unexpected is the Poisson distribution of the variable 'diagnosis asthma': the percentage of diagnosed asthma occurrences in the city is not significantly higher than in the rural areas. Even more surprising is the distribution of the variable 'asthma symptoms during the last 12 months'; the

Table 11.2 Age group 12–14 years

Variable	p	Degrees of freedom	χ^2	H_0
Diagnosis hay fever	0.092	17	36.99	rejected
Hay fever symptoms	0.219	20	123.07	rejected
Diagnosis asthma	0.040	4	1.24	not rejected
Asthma symptoms	0.116	19	53.36	rejected
Diagnosis endogenous eczema	0.055	13	27.91	rejected
Endogenous eczema symptoms	0.076	16	22.32	not rejected

H_0: the number of children (12–14 years) with these illness or symptoms is Poisson distributed over the ZIP-code areas.

urban area shows no significantly higher rates than the climatically more exposed areas.

The next stage at the ZIP-code level will be to search for possible causes at this level. We assume that the genetic circumstances are equal over the whole research area, so we are looking for environmental and possibly social causes for the differences in the distributions. In Greifswald, we can also expect a differentiation in the distribution of the variables, so it is necessary to take a more detailed look, which we will do at an address level.

When the epidemiological data in the ZIP-code regions have to be compared with environmental data, the problem of data integration and polygon overlay appears. Environmentally defined regions will seldom match with ZIP-code regions. Flowerdew and Green (1994) solve this problem by combining the zones of the different regions (called target and source zones) to intersection zones and introducing additional information into the areal interpolation process.

Stage two of the ERIS project will concentrate more on statistics and spatial analysis. For the ZIP-code level, methods for small areas are appropriate. Cuzick and Elliot (1992) consider any region containing fewer than about 20 cases of disease as a small region. Of course, the city of Greifswald shows more than 20 disease cases for most categories of disease, but this is an exception in the study area. It is important to recognise possible misleading results, coming from the difference in population density between the city and the surrounding area and to use smoothing techniques to correct for these extreme differences. Empirical Bayesian estimates may be used to smooth the disease rates over the map. These are a compromise between the classical standardised ratios and the overall mean for the whole map (Clayton and Kaldor, 1987; see also López-Abente and Braga, chapters 10 and 9, respectively). Bayes methods are based upon the number of cases in the research area: estimated rates, based on a large number of cases, tend to differ little from the standardised rates. On the other hand, estimates based upon only a few cases are drawn towards the overall mean of the research area (Clayton and Bernardinelli, 1992).

11.4.2 Spatial analysis at an address level in the city of Greifswald

The results of the initial analysis at the ZIP-code level show that the number of cases with diagnosis hay fever, hay fever symptoms during the last 12 months and diagnosis endogenous eczema is more than expected. In these cases a further analysis at address level is definitely worthwhile, but also the other variables, with little or no deviation from the expected count in the whole city, can show a non-random spatial pattern within the urban region.

The point data have the following properties: for each child, the complete address is known, so we can attach a pair of xy-coordinates to every child. There are about 80 attribute values known at each point, both information about the different diseases (allergies) and information about environment and lifestyle. The information about the diseases is mainly bivariate: the individuals have the disease ('cases') or they have not ('controls') (Gatrell and Bailey, 1996).

Because of data confidentiality, we cannot visualise the raw point data as they stand. We have to aggregate the data to small areas. If we do not wish to divide the research region into small administrative areas, we can work with buffer zones, for example around streets with a high level of waste gases, if possible in combination with inner-city wind patterns.

The epidemiologists constructed two broad hypotheses as a starting point for the spatial analysis on this level. First, that there is an obvious relationship between the volume of vehicles, especially lorries, and different allergy symptoms. Second, that there could be a relationship between social status and different allergy symptoms.

We first want to know if there is any clustering in one of the six symptoms. A variety of methods exist to detect clusters and clustering in a point map (see, for example, Gatrell and Bailey, 1996, and Kulldorff, chapter 4). Wartenberg and Greenberg (1990) describe a strategy to select an appropriate method of cluster detection. First, the selection of the data type: the *location* of an event, the *distance* between all pairs of events, the *nearest-neighbour* distance between events, or the distance to a *fixed point*. Second, selection of a transformation type: no transformation, ranks, inverse of the distances or some other type of transformation. Third, selection of a summary statistic: for example counts per cell, mean distances. Fourth, selection of a method to generate the reference distribution with which the observed distribution has to be compared. Wartenberg and Greenberg compare different methods with the help of simulations.

After finding an eventual spatial pattern in the data, this pattern has to be compared with the spatial pattern of possible causal factors. Correlation, covariance and regression methods are often used to detect relations between variables. Since disease data often have a binomial character, logistic regression methods will often be useful.

11.5 Further development of ERIS

Most of the epidemiologists are not GIS specialists; nevertheless they are interested in the methods offered by GIS. An application has to be fitted to the needs and requirements of the users, and with this in mind, they must be involved in the developing process, to guarantee the quality of the application. Without an interdisciplinary approach, the development of a good application will be very difficult.

The ERIS project incorporates the software AtlasGIS, which is a relatively simple PC GIS, often used in Germany. The application is being developed using this software, since it runs on a PC, is modest in cost, easy to learn and can read different attribute database formats. AtlasGIS also has some interfaces to other spatial data formats.

What analytical features are required in such an application? In the first place it must be possible to visualise the data in different ways: not only as maps, but also as graphics and diagrams. Secondly, it must include modules to carry out spatial analysis. These include standard GIS-functions for generating buffer zones around point, line and area objects, functions for measuring and counting, Boolean logic and overlay operations, and standard statistical functions. Statistical functions, needed for spatial analysis, are rare in much GIS-software and not present in AtlasGIS. Functions required in an epidemiological application include, for example, proof of distributions (Poisson, normal), data reduction (factor analysis), different methods of spatial cluster analysis, correlation, and regression. Some of these functions could be coupled to the application via a conventional statistical package. But other problems arise in the field of typical spatial statistics, like spatial cluster analysis and detection. These analysis methods must either be programmed separately or integrated in the application through links with spatial statistical software (see Gatrell and Bailey, 1996; Fotheringham and Rogerson, 1994). Linking the ERIS application to spatial statistical software will be a part of the next ERIS project (ERIS2), which started in February 1997. An important part of the application will be an extended help function. Complicated spatial statistics are not easy to use and an understandable step-by-step explanation of the data constraints and methods is necessary.

The last but certainly not the least important part of this chapter involves the interaction between the human and the computer, or the man–machine interaction. The user interface is the key to the system. An application is only useful, when, in this case the epidemiologist, is able to make use of it. Firestone et al. (1996) suggest that effective user interfaces optimise interactions between the operator and the workstation and that interfaces optimise interactions by minimising and making intuitive required inputs, and providing clear feedback on the progress and status of operations. Van der Veen (in this volume) adds the point that the user interface should not allow attempts that would damage the consistency of the database. During the development stage of the user interface, contact must be maintained with the later users. This is just as important as the involvement at the application development stage. The requirements of the users must be the most important information source for the designer. The opportunity for evaluation in order to check whether the system is meeting the requirements of the user is necessary (Medyckyj-Scott, 1993). If the user interface allows an easy interactive admittance to the system, the GIS application can be a useful tool in epidemiology.

11.6 Conclusions

The use of spatial analysis in health research projects is growing, but there are still important issues to be tackled. After eight months of research in the field of GIS and health, we have made some progess in digitising basic map coverages, importing and georeferencing the ISAAC data, visualising queries on different aggregation levels and making some simple statistical analysis. Yet we know that such health data often are incomplete, do not cover the whole area or are quantitatively difficult to handle. In all these cases, spatial statistical analysis is quite complicated, but represents the only way to make significant statements about spatial patterns in epidemiological data.

Nevertheless, the subject of GIS and health holds great promise. Visualisation is a first step, but will make things clearer. Subsequent, well-considered statistical tests can then provide new understanding of associations between epidemiological and environmental phenomena. In Germany, the use of GIS in epidemiology is still rare. Making spatial statistical techniques available for epidemiologists in GIS applications can help to make progress in epidemiological research. The ERIS project is still in its infancy, but we can see already the advantages of visualisation, spatial analysis and modelling in epidemiology.

References

ASHER M. I., KEIL U., ANDERSON H. R., BEASLEY R., CRANE J., MARTINEZ F., MITCHELL E. A., PEARCE N., SIBBALD B., STEWART A. W., STRACHAN D.,

WEILAND S. K. and WILLIAMS H. C. (1995) International study of asthma and allergies in childhood (ISAAC): rationale and methods. *European Respiratory Journal*, **8**, 483–491.
BLALOCK H. (1981) *Social Statistics*. McGraw-Hill, Singapore.
BORTZ J. (1993) *Statistik für Sozialwissenschaftler, 4.überarb. Auflage*. Springer, Berlin.
BROWN P. J. B., BATEY P. W. J., HIRSCHFIELD A. and MARSDEN J. (1990) Poisson chi square mapping, GIS and geodemographic analysis: the spatial and aspatial analysis of relatively rare conditions. *Working paper 18*, URPERRL, Liverpool.
CLAYTON D. and BERNARDINELLI L. (1992) Bayesian methods for mapping disease risks. Pages 205–220 in Elliot P., Cuzick, J., English D. and Stern R. (eds) *Geographical and Environmental Epidemiology: Methods for Small Area Studies*. Oxford University Press, Oxford.
CLAYTON D. and KALDOR J. (1987) Empirical Bayes estimates of age-standardized relative risks for use in disease mapping. *Biometrics*, **43**, 671–681.
CUZICK J. and ELLIOT P. (1992) Small-area studies: purpose and methods. Pages 14–21 in Elliot P., Cuzick J., English D. and Stern R. (eds) *Geographical and Environmental Epidemiology: Methods for Small Area Studies*, Oxford University Press, Oxford.
DIGGLE P. J., GATRELL A. C. and LOVETT A. A. (1990) Modelling the prevalence of cancer of the larynx in part of Lancashire: a new methodology for spatial epidemiology. Pages 34–47 in Thomas R. W. (ed.) *Spatial Epidemiology*, Pion, London.
EDWARDS B. (1987) *Drawing on the Artist Within*. Collins, London.
FIRESTONE L., RUPERT S., OLSON J. and MUELLER W. (1996) Automated feature extraction: the key to future productivity. *Photogrammetric Engineering and Remote Sensing*, **6**.
FLOWERDEW R. and GREEN M. (1994) Areal interpolation and types of data. Pages 121–145 in Fotheringham S. and Rogerson P. (eds) *Spatial Analysis and GIS*, Taylor and Francis, London.
FOTHERINGHAM S. and ROGERSON P. (eds) (1994) *Spatial Analysis and GIS*, Taylor and Francis, London.
GATRELL A. C. (1994) Density estimation and point patterns. Pages 65–75 in Hearnshaw H. and Unwin D. J. (eds) *Visualization in Geographical Information Systems*, Wiley, Chichester.
GATRELL A. C. and BAILEY T. C. (1996) Interactive spatial data analysis in medical geography. *Soc. Sci. Med.*, **42**, 843–855.
GATRELL A. C., BAILEY T. C., DIGGLE P. J. and ROWLINGSON B. S. (1996) Spatial point pattern analysis and its application in geographical epidemiology. *Transactions, Institute of British Geographers*, NS, **21**, 256–274.
HEARNSHAW H. (1994) Psychology and displays in GIS. Pages 193–199 in Hearnshaw H. and Unwin D. J. (eds) *Visualization in Geographical Information Systems*, Wiley, Chichester.
HEARNSHAW H. and UNWIN D. J. (eds) (1994) *Visualization in Geographical Information Systems*, Wiley, Chichester.
JACQUEZ G. M. and WALLER L. A. (1996) The effect of uncertain locations on disease cluster statistics. *Proceedings of the Second International Symposium on Spatial Accuracy Assessment*, Fort Collins, Colorado.
MEDYCKYJ-SCOTT D. (1993) Designing Geographical Information Systems for use. Pages 87–100 in Medyckyj-Scott D. and Hearnshaw H. M. (eds) *Human Factors in Geographical Information Systems*, Belhaven Press, London.
MÖHNER M. and STABENOW R. (1994) Kartographische Darstellung epidemiologischer Sachverhalte und Methoden zur Clustererkennung. Pages 679–692 in Heinemann (ed.) *Epidemiologische Arbeitsmethoden*, Fischer, Jena.
STERN R. M. (1995) Environment and health data in Europe as a tool for risk management: needs, uses and strategies. Pages 3–24 in de Lepper M. J. C., Scholten H. J. and Stern

R. M. (eds) *The Added Value of Geographical Information Systems in Public and Environmental Health*, Kluwer Academic Publishers, Dordrecht, 3–24.

VAN DER VEEN A., FEHR R. and PRÄTOR K. (1996) Exchange of spatial data for environmental health information management (EHIM). Paper prepared for the Helsinki GISDATA meeting, May 1996.

WALLER L. A. and JACQUEZ G. M. (1995) Disease models implicit in statistical tests of disease clustering. *Epidemiology*, **6**, 584–590.

WARTENBERG D. and GREENBERG M. (1990) Space-time models for the detection of clusters of diseases. Pages 17–34 in Thomas R. W. (ed.) *Spatial Epidemiology*, Pion, London.

WEBSTER R. and OLIVER M. A. (1990) *Statistical Methods in Soil and Land Resource Survey*. Oxford University Press, Oxford.

WESTLAKE A. (1995) Strategies for the use of geography in epidemiological analysis. Pages 135–144 in de Lepper M. J. C., Scholten H. J. and Stern R. M. (eds) *The Added Value of Geographical Information Systems in Public and Environmental Health*, Kluwer Academic Publishers, Dordrecht, 135–144.

WOOD M. (1994) The traditional map as a visualization technique. Pages 9–17 in Hearnshaw H. and Unwin D. J. (eds) *Visualization in Geographical Information Systems*, Wiley, Chichester, 9–17.

CHAPTER TWELVE

Problems and Possibilities in the Use of Cancer Data by GIS – Experience in Finland

LYLY TEPPO

12.1 Introduction

Various environmental factors are important in the causation of human cancer. The role of smoking, diet, sexual habits and certain infections are usually analysed at the level of individuals, since they are properties of those individuals. The effects of some important exogenous exposures, such as ultraviolet and ionising radiation, and ambient air and water pollution, depend on where people live, where they work and as a result analyses call for geographical thinking.

In this chapter I first discuss the nature and quality of data available on the occurrence of cancer and then review studies in which different geographical approaches have been applied in epidemiological cancer research in Finland.

12.2 Data on the occurrence of cancer

Detailed data on the occurrence of cancer in a given population are based on either new cancer cases (incidence) or cancer deaths (mortality). Incidence information is collected by population-based cancer registries. The few country-wide registration systems include those in the Nordic countries (Denmark, Finland, Iceland, Norway, Sweden), UK and Canada. Many registries cover only one region or just one town. In fact, some country-wide registrations are based on a number of regional registries.

In a cancer registry information is stored on all cancers diagnosed in the target population. The data collection methods vary, and so do the items recorded. Most registries include identification data for the patients (date of birth, person number if available, address or place of residence), information about the tumour (date of diagnosis, primary site, method of verification of the diagnosis, histological type, stage, treatment) and follow-up data (date and cause of death, date of emigration).

These data enable the production of routine cancer statistics and various exercises in the field of descriptive epidemiology, such as incidence trends and cancer maps. Data on the level of individuals are essential in record linkage procedures in which data from other sources (exposure data) are combined with cancer registry data. If unique person numbers are available the record linkage is easy and can be computerised.

From the GIS point of view, data on patient residence are crucial. The accuracy of this information varies between registries. Sometimes one records the residential municipality which permits geographical analyses in which the municipality is the basic unit. More sophisticated and detailed methodology can be applied if individual patients can be located more precisely. In some registries the street address of the patient is available. Use of addresses is often very laborious. In Finland the coordinates (by 10 m) of all houses are available which enables the construction of *ad hoc* areas and populations irrespective of administrative borders. It is easy to comprehend the usefulness of this data set in cancer research: cancer risks within any geographically defined (exposed) population can be evaluated. Examples are given towards the end of this chapter.

Mortality data are available from almost every population in the world. The individuals with cancer as the cause of death can be identified, and cancer mortality data by primary site created. The address (or other information of the place of residence) of the patient at the time of death is usually recorded, and can be used for different analyses of a geographical nature.

12.3 Problems in the quality of cancer data

If cancer incidence or mortality data cover all cancer patients and only genuine cancer patients, and if the data on individual cancer patients are correct and detailed enough, their use for epidemiological research and for various geographical analyses can yield a lot of interesting and valuable data on the background factors of cancer causation and give clues for primary prevention. However, in reality various problems may be encountered, some of which are of minor importance but some of which are serious enough to merit attention by those who use the data (Saxén and Hakama, 1975).

12.3.1 Diagnosis of cancer

In many instances the diagnosis of cancer in a given individual (patient) is clear-cut and unequivocal. The symptoms of the patient, various clinical observations including imaging, endoscopies and operations lead to microscopical investigation in which a pathologist interprets the findings. In some 90% of cancer cases (in most developed countries) the diagnosis of cancer is based on microscopical investigation of either a tissue specimen or a cytological specimen. For instance, the morphological diagnosis of skin cancer, breast cancer, oral cancer, bladder cancer and colon cancer is quite straightforward and fully reproducible by different pathologists.

But sometimes the morphological diagnosis of cancer is not that clear. Difficulties in the interpretation of tissue specimens can be encountered, for example, in lymph node pathology and in soft-tissue lesions (malignant tumour or not). Not infrequently, the tissue specimen submitted to the pathologist is very small or compressed or contains only necrotic tissue so that no proper diagnosis can be reached.

The concept of cancer is sometimes difficult to apply to morphological diagnosis (Saxén, 1979). What is cancer really? There are tumours of borderline malignancy (the clinical behaviour is not estimable), especially in the ovaries. And there is the well-known problem of differentiating between two pre-clinical lesions of the cervix uteri, dysplasia gravis and carcinoma *in situ*. The former is not cancer, whereas the latter can already be considered as an early stage of cancer (and recorded as such by cancer registries). Similar problem areas – cancer or not – are numerous, further examples being small clear-cell tumours of the kidney, papillomas of the urinary tract and glial proliferations of the brain. In these instances the final diagnosis is subject to inter-observer variation and leads to some uncertainty in cancer statistics and other patient-based analyses in which detailed stratifications of the material are needed.

Further, microscopical specimens are not always available for study. This is true especially in lesions of the internal organs which may be difficult to reach for diagnostic purposes (biopsy). In these instances the diagnosis must be based on imaging techniques (for example, lung cancer, brain cancer, bone cancer), on some laboratory tests (multiple myeloma) or just on clinical findings (cancer in the abdominal organs). In Finland, the proportion of cases with microscopical verification of the diagnosis was (in 1993) 74% for pancreas cancer, 83% for brain tumours and 89% for lung cancer – the average for all cancers taken together was 94%.

Without morphological verification the certainty of cancer diagnosis varies substantially. If metastases are found, for example, by X-ray, the diagnosis of cancer is in most instances clear, but its origin (primary site) may remain unknown. This, of course, may be the case also when the existence of a metastasis has been verified by biopsy and histological investigation. In the Finnish Cancer Registry the primary site of cancer is unknown for about 3% of all cases. If the diagnosis is based on clinical examination only, false cancer diagnoses can be made, recorded as cancers in various instances, and used misleadingly as cancers in scientific analyses.

Even if there may be problems in the diagnosis of cancer this is true only for a minor fraction of all cases. Moreover, the diagnostic criteria of cancer are much more well-defined and reproducible than those of many other diseases (such as rheumatoid arthritis, bronchial asthma, ischaemic heart disease or mental disorders).

12.3.2 Coverage of cancer statistics

The usefulness of cancer statistics and any data based on individual cancer patients depends on the coverage: are all cancer patients included? Problems may be encountered at the diagnostic phase (diagnosis is never made) or in data collection (patient is never recorded at a cancer registry). Undoubtedly, numerous cancers remain undetected during the lifetime of the individual. Autopsy studies have shown that this is especially true in old individuals and in cancers of the internal organs, such as cancers of the kidney and ovary. These tumours may be totally symptomless, or the symptoms are general and do not readily lead to cancer diagnosis (fatigue, fever, weight loss, etc.). Knowing that it takes several years for a small cancerous focus to grow and surface as a clinically important disease it is self-evident that there always are a lot of individuals in the population with small undiagnosed cancers, part of which will also remain undetected. Cancer incidence figures produced by cancer registries refer, of course, only to diagnosed cancers.

When cancer trends or geographical or other variations in cancer occurrence are considered, it is important to notice that if the proportion of undiagnosed cancer is stable in time and space and by age, the data available are still fully useful. But this is not always the case. Better diagnostic methods are developed all the time, and smaller and smaller lesions can be detected, resulting in an increase, over time, in the incidence without a real (biological) increase in the occurrence of cancer. Diagnostic standards may vary between different countries and different parts of the country which can result in misleading patterns. The coverage also depends on age: an individual with undiagnosed cancer is likely to be old. Cancer at a young age will appear clinically after some time while old individuals often die of other diseases and their cancers may thus remain undiagnosed. This is why one might be willing to exclude in some analyses the oldest age-groups, for example, those over 70 or 75 years.

In Finland, registration of chronic leukaemias and some other haematological disorders have been slower than that of solid tumours; reporting to the Registry is often delayed and the first information is received from death certificates (Teppo *et al.*, 1994). The result is that for these cancers there is certain undercoverage in the latest reported period, which disappears in a couple of years.

As a result of such problems, variation over time, and between different countries and regions, in the coverage of cancer, diagnoses and registration must be taken into account in many studies. Analyses could perhaps be restricted to areas with equal standards in data production, or periods in which marked diagnostic changes took place could be excluded. But here too it is reassuring to notice that in terms of coverage, cancer data are usually much better in quality than those for most other diseases.

12.3.3 Relevance of data on residence

Even if data on the place of residence at the time of the diagnosis of cancer were accurate and complete, their relevance can be questioned. No problems are likely to occur when this information is used for administrative purposes. For instance, cross-sectional analyses in which residence and cancer data refer to the same point of time are totally appropriate for planning purposes. But those who are interested in risk factors and exposure distributions or have other ideas about cancer causation in their minds think, consciously or unconsciously, that the place of residence could have something to do, directly or indirectly, with the causes of cancer involved. This calls for longitudinal studies.

Cancer always has a long latency period from the onset of the process to clinical disease. Excluding the rare cases of cancer among infants, this latency may vary from some 10 years (for example, irradiation-induced leukaemia) up to more than 40 years (for example, asbestos-related pleural mesothelioma). In many instances the exposure must last a lengthy period in order to be able to induce and promote the growth of cancer. This means that the residence of the patient at the time of diagnosis is not very relevant in terms of cancer causation unless the individual has lived in the same place for decades. Since the latency periods vary substantially by cancer type and also from each individual cancer to another, it is not possible to define the correct point of time prior to cancer diagnosis which would be the most relevant in terms of cancer causation. A complete exposure history would, in the case of residence, mean information of the addresses of all houses in which the individual has lived, say, 10–50 years prior to the diagnosis. This information is seldom available. This is likely to obscure the relationship between residence and risk of cancer.

Another thing is that people do not stay home day and night. They go to work and have leisure-time activities sometimes rather far from their homes, not infrequently in another municipality. Such issues are discussed by Löytönen in chapter 7.

Researchers have accepted the fact that occupation at the time of cancer diagnosis is not relevant in terms of occupational risk factors in cancer, unless it has been stable for a lengthy period of time. But occupation may be an appropriate surrogate for social class and education and thus indirectly reflect certain risk factors related to the way of life. Similarly, one should think that even if the physical environment of the patient's home is not always an appropriate parameter as such, it can sometimes be used as an indicator of some other characteristics interesting from the point of risk of cancer. However, whenever possible, a certain latency should be allowed in the analyses of the relationship between residential data and occurrence of cancer as is always done in classical epidemiological (cohort or case-control) approaches.

12.4 Cancer incidence or mortality?

In principle, incidence data are better than mortality data for most research purposes. They describe directly the occurrence of clinical cancer, whereas in mortality data the survival experience of cancer patients is also involved. Since survival rates vary substantially between populations, variation in mortality rates does not always describe adequately the variation in cancer occurrence. Only for cancers with poor prognosis everywhere (such as cancers of the lung and stomach) are mortality data an acceptable substitute for incidence data.

Certain important cancers have a rather favourable outcome; for example, less than 25% of breast cancer patients die of their cancer, and for lip cancer this proportion is almost nil. In these instances the mortality rates – even if they would be proportional to incidence rates – are much smaller and their reliability thus lower than in the case of incidence data.

Unfortunately, incidence data are available only for a limited number of areas in the world. Moreover, cancer incidence data can be of poor quality and their use in research thus questionable. The mortality/incidence ratio has been used as an indicator of the quality of cancer registration in a given area. If this ratio exceeds unity one should consider the possibility of under-registration of incident cancer cases.

A high autopsy rate has often been considered as a guarantee of reliable incidence (and mortality) rates. Of course, a number of undiagnosed cancers are detected and in some instances the clinical diagnosis of cancer and its exact site of origin can be confirmed. But in the case of widespread cancer for which the coding problem (primary site) is greatest, the autopsy does not help too much; for example, if a large part of the abdominal cavity is occupied by tumour tissue, it is not possible to define its origin. The autopsy rate varies by age within a given population. Young and middle-aged patients are autopsied more often than older ones for which the need of further information would be greatest. If the autopsy rate varies between regions, problems in the spatial comparability of data may occur.

Researchers who are able to use cancer registry-based incidence data are in an exceptionally good position, since despite various shortcomings cancer incidence data are superb with improvements in them also taking place continuously.

12.5 Examples of geographical approaches to utilise cancer incidence data in Finland

In the Finnish Cancer Registry, two types of residential data are available: municipality (or any larger area composed of municipalities) at the time of diagnosis of cancer (1 January that year), and coordinates of the houses in which the patient had been living prior to cancer diagnosis.

Several simple area-based cancer maps have been produced, based on municipality data. There are some 450 municipalities in Finland (current population 5.1 million). Larger units have been the 12 provinces of the country or the 22 health care districts. For many cancers this approach has clearly demonstrated a north–south (or north-east–south-west) gradient in the incidence within Finland (Teppo et al., 1975), or substantial and interesting variations in the incidence within the Nordic countries (Møller Jensen et al., 1988). Cancers that are known to be more common in upper social classes (for example, those of the breast, colon and kidney) are more common in southern and south-western parts of Finland (Hakama et al., 1979); see Figure 12.1a. Accordingly, male lung cancer, which is more common in lower social classes, shows higher incidence rates in northern and north-eastern parts of the country where smoking has traditionally been more prevalent than elsewhere (Figure 12.1b).

A lot of information on the standard of living, economy, occupational distribution and industry is available for each individual municipality. In an earlier ecological study the municipalities were first divided into three to five classes according to the numerical value (percentages, means or continuing variables) of several background variables (Teppo et al., 1980). Then age-specific numbers of incident cancer cases within each group, and corresponding population data, were used for cancer incidence calculations. In this way the variation between municipalities (both in terms of background factor in question and, consequently, the risk of cancer) could be demonstrated more clearly than using administrative areas which are always heterogeneous and include both well-to-do towns and more or less underdeveloped remote rural areas. An example of this approach is given in Figure 12.2.

A smoothing methodology using municipality-based cancer incidence data was applied in the Finnish Cancer Atlas study (Pukkala et al., 1987). The colour of each dot on the map was defined on the basis of the weighted mean of the cancer incidence rates in the nearby municipalities, the weights being indirectly proportional to the distance between the municipality centre and the dot to be coloured, and directly proportional to the population size of the municipality. This type of presentation, similar to that discussed elsewhere in this book, aims at hiding unstable rates of individual municipalities; the general patterns within the country emerge rather effectively. Sparsely populated areas in northern Finland are a problem in this (and in any) approach.

A similar approach is being applied in the Cancer Atlas of Nothern Europe project which covers Finland, Sweden, Norway, Iceland, Denmark, Germany,

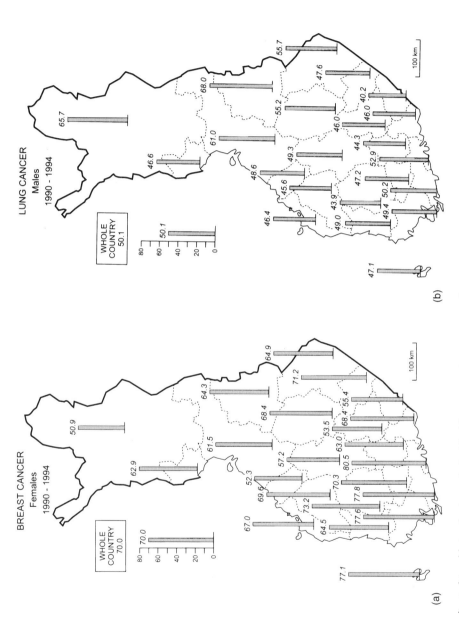

Figure 12.1 Age-adjusted (world standard population) incidence rates (per 100 000) of cancers of the (a) breast (females) and (b) lung (males) in Finland in 1990–1994, by health care district.

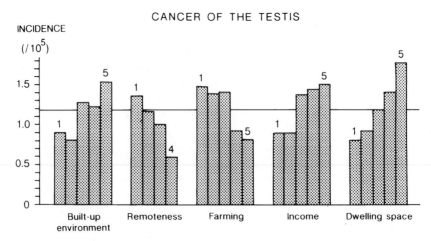

Figure 12.2 Age-adjusted (world standard population) incidence rates (per 100 000) of testis cancer in four or five groups of municipalities in Finland in 1955–1974 stratified by different variables describing the well-being of the municipality (Teppo et al., 1980). Horizontal line: rate for the total population. *Built-up environment*: percentage of population living in urban-like centres in 1970. *Remoteness*: distance between the municipality centre and centre of the local economical area. *Farming*: percentage of population in farming and forestry in 1950. *Income*: average income per inhabitant in 1968. *Dwelling space*: mean dwelling space per inhabitant. 1 = numerical value of the variable lowest, 4 or 5 = numerical value highest. Notice that for remoteness and farming, the most well-to-do municipalities constitute group 1, for other variables they are included in group 5.

Poland, Lithuania, Latvia, Estonia, Belarus and the western part of Russia. The problems that have emerged include the uneven size of the basic areal units in different countries and problems in the availability of population data in some eastern European countries.

Cancer incidence data by municipalities in Finland were used for the evaluation of the effects of the Chernobyl accident (in 1986) on the risk of leukaemia in children 0–14 years of age (Auvinen et al., 1994). Expected numbers of leukaemia (by sex and age) in 1986–1988 and 1989–1992 were calculated for each municipality on the basis of incidence data from 1976–1985, and pooled as quintiles of exposure (fallout exposures were measured for all municipalities separately using continuously moving cars). Standardised incidence ratios of childhood leukaemia and relative excess risk of childhood leukaemia per mSv were assessed. The relative excess risk in 1989–1992 was not significantly different from zero (7% per mSv), and an important increase in childhood leukaemia could thus be excluded.

The relationship between exposure to mutagenic drinking water and cancers of the gastrointestinal and urinary tract was studied by Koivusalo et al. (1994). Past exposure to drinking water mutagenicity was assessed in 56 selected urban municipalities for the years 1955 and 1970. The numbers of cases of cancer in these municipalities were derived from the Finnish Cancer Registry for 1967–1976 and 1977–1986. Age, sex, social class, urban living, and time period were taken into account in the Poisson regression analysis. In 34 municipalities using chlorinated surface water this exposure indicated a significantly elevated relative risk of 1.2 for bladder cancer and 1.2–1.4 for kidney cancer compared with municipalities where nonmutagenic drinking water was consumed. No excess risks could be demonstrated for cancers of the stomach, colon and rectum.

Residential coordinates have been used in several occasions when cancer risks within various *ad hoc* areas have been analysed (see Pukkala, 1992). In the study of the relationship between exposure to magnetic fields generated by high-voltage power lines and risk of different childhood cancers (Verkasalo et al., 1993), the corridors around the power lines were first computerised (the coordinates of the towers were known). Thereafter, the houses within these corridors and people who had ever lived in these houses were defined, and, finally, cancer cases identified from the Cancer Registry. It was possible to analyse the effect of this exposure in terms of risk of childhood cancer using several ways to calculate or measure the exposure to magnetic field. The results were largely negative. A statistically significant excess of nervous system tumours was found in boys (but not in girls) who were exposed to magnetic fields of >0.20 mT or cumulative exposure of >0.40 mT years. The project is ongoing and considers now different adult cancers with some 10 000 observed cases within the exposed area.

Another study concentrated on the evaluation of the risk of lung cancer in 33 sub-areas of the City of Helsinki, and separately among those residing along the main streets (Pönkä et al., 1993). House coordinates were used to define the different target populations, and the lung cancer cases were again identified from the Cancer Registry. Data were available on the level of education of people and on the average levels of certain air pollutants such as SO_2 and NO_2. The main result was that variation in the level of education (which is a surrogate of smoking habits) explained most of the variation in the lung cancer risk both between subareas (standardised incidence ratios 0.56–1.56 for males, 0.29–3.17 for females) and between people living close to main streets in different parts of the town (SIRs 0.39–1.31 for males, 0.24–1.51 for females).

A nested case-control approach was used in the evaluation of the relationship between indoor radon exposure and lung cancer risk (Auvinen et al., 1996). Lung cancer cases (1973), and controls matched by sex and age, were drawn from a cohort of all those who had been residing in the same one-family house for 19 years (1967–1985); lung cancers were identified through record linkage to the Finnish Cancer Registry. The basic population had been defined using the register of house coordinates. Based on information given in questionnaires, and 12-month measurements of radon concentration, the odds ratio was 1.11 (95% confidence interval 0.94–1.31) per 2.7 pCi/l (100 Bq m^{-3}) after adjustment for smoking habits. There was no statistically significant risk associated with radon, but the risk estimate was very close to that found in other studies.

Another approach to study the association between indoor radon and lung cancer was applied by Ruosteenoja et al. (1996). From the Finnish Cancer Registry they identified all male lung cancer cases diagnosed in 1980–1985 in nineteen rural municipalities in southern Finland known to have high levels of radon exposure (and very low levels of air pollution). The 238 cases and their 495 population-based controls were interviewed (for example, for their smoking habits and occupational histories), and the radon exposures were measured in all houses in which the cases and controls had been living for more than one year in 1950–1975. The authors concluded that the effect of radon on lung cancer risk was only suggestive and not statistically significant.

In the analysis of the risk of cancer (especially leukaemia) in the surroundings of an oil refinery on the southern coast of Finland, and in further studies of the relationship between certain characteristics of drinking water (such as radon and arsenic concentrations) and cancer, information on the house coordinates is being used in order to define the study populations. In a multinational study on the clustering of childhood cancer (EUROCLUS), a similar approach is applied (Alexander et al., 1996).

12.6 Concluding remarks

There are several favourable issues in terms of analysing cancer risks and the effects of different carcinogenic exposures using geographical data. The definition of cancer is rather clear-cut which enables the production of meaningful and reliable incidence data to be used in research. The existence of cancer registration systems gives exceptional opportunities for researchers to use data collected during decades; in other words, one can avoid the time-consuming and expensive collection of basic data during each individual study project.

But several details must be remembered and taken into consideration. There may be uncertainty in the diagnosis of some cancers, in the coverage of registration systems and especially in the accuracy of mortality data. The latency between the onset of a carcinogenic process and clinical cancer is vital. And finally, many cancers are rare, and to increase the power of the study and to get meaningful results one has to prolong the study period which may diminish the possibility of finding associations that are scientifically relevant and useful at different levels of cancer control. Another way would be to expand the study over national borders which of course may lessen the comparability of data but on the other hand will markedly increase variation in exposure levels.

Production of simple maps may in some instances result in interesting and useful findings in terms of cancer causation, and maps often serve as a means of generating hypotheses. But the main use of individual-based cancer data combined with information on the residence is in analytical epidemiological studies in which the relationship between different exposures and occurrence of cancer is evaluated using a wide variation of designs and adequate modern biostatistical methodology.

Acknowledgment

The assistance received from Eero Pukkala, PhD, from the Finnish Cancer Registry in the preparation of the manuscript is greatly appreciated.

References

ALEXANDER F., WRAY N., BOYLE P. et al. (1996) Clustering of childhood leukaemia: a European study in progress. *J Epid Biostat*, **1**, 13–22.

AUVINEN A., HAKAMA M. and ARVELA A. (1994) Fallout from Chernobyl and incidence of childhood leukaemia in Finland. *Brit Med J*, **309**, 151–154.

AUVINEN A., MÄKELÄINEN I., HAKAMA M. et al. (1998) Indoor radon exposure and risk of lung cancer: a nested case-control study in Finland. *J Nat Cancer Inst*, **90**, 401–402.

HAKAMA M., SOINI I., KUOSMA E. et al. (1979) Breast cancer incidence – geographical correlations in Finland. *Int J Epid*, **8**, 33–40.

KOIVUSALO M., JAAKKOLA J. J. K. and VARTIAINEN T. (1994) Drinking water mutagenicity and gastrointestinal and urinary tract cancers: an ecological study in Finland. *Am J Pub Health*, **84**, 1223–1228.

MØLLER JENSEN O., CARSTENSEN B., GLATTRE E. *et al*. 1988. *Atlas of Cancer Incidence in the Nordic Countries. A Collaborative Study of the Five Nordic Countries*. Nordic Cancer Union, Helsinki.

PÖNKÄ A., PUKKALA E. and HAKULINEN T. (1993) Lung cancer and ambient air pollution in Helsinki. *Environm Int*, **19**, 221–231.

PUKKALA E. (1992) Use of record linkage in small-area studies. Pages 125–131 in Elliot P., Cuzick J., English D. and Stern R. (eds) *Geographical Epidemiology*, Oxford University Press, Oxford.

PUKKALA E., GUSTAVSON N. and TEPPO L. (1987) *Atlas of Cancer Incidence in Finland 1953–82*. Cancer Society of Finland publications no. 37. Finnish Cancer Registry, Helsinki.

RUOSTEENOJA E., MÄKELÄINEN I., RYTÖMAA T., HAKULINEN T. and HAKAMA M. (1996) Radon and lung cancer in Finland. *Health Phys*, **71**, 185–189.

SAXÉN E. (1979) Histopathology in cancer epidemiology. The Maude Abbott Lecture. *Path Ann*, **14**, 203–217.

SAXÉN E. and HAKAMA M. (1975) Trends in cancer incidence – facts or fallacy. Studies in Finland. Pages 168–174 in Bucalossi P., Veronesi U. and Cascinelli N. (eds) *Proceedings, XI International Cancer Congress*, Florence, 20–26 October 1974, vol 3. Excerpta Medica, Amsterdam.

TEPPO L., HAKAMA M., HAKULINEN M. *et al*. (1975) Cancer in Finland 1953–1970. Incidence, mortality, prevalence. *Acta Path Microbiol Scand* Sect A Suppl 252.

TEPPO L., PUKKALA E., HAKAMA M. *et al*. (1980) Way of life and cancer incidence in Finland. A municipality-based ecological analysis. *Scand J Soc Med Suppl*, **19**.

TEPPO L., PUKKALA E. and LEHTONEN M. (1994) Data quality and quality control of a population-based cancer registry. Experience in Finland. *Acta Oncol*, **33**, 365–369.

VERKASALO P. K., PUKKALA E., HONGISTO M. Y. *et al*. (1993) Risk of cancer in Finnish children living close to power lines. *Brit Med J*, **307**, 895–899.

CHAPTER THIRTEEN

GIS in Public Health

PAUL WILKINSON, CHRISTOPHER GRUNDY,
MEGAN LANDON and SIMON STEVENSON

13.1 Introduction

Where we live and work determines many factors that influence our health, including, to a large degree, the treatment we receive from health services. Geographical studies therefore have an important role in many aspects of public health and health service planning. No list can be comprehensive, but some of the main types of geographical analysis used in this context include:

1. Examination of disease rates and other health statistics by geographical area to assess the health of the population.
2. Examination of variation in health and use of health services as a comparative approach to needs assessment and resource allocation.
3. Examination of time trends in disease at a local level.
4. Analysis of the spatial distribution of health care facilities and referral patterns to aid decisions about optimal location of health services.
5. Studies of variation in health treatments and outcome for planning the development of health services.
6. Studies of health and health promotion interventions at community level.
7. Disease surveillance, for example of communicable diseases, congenital malformations.
8. Analytical epidemiological studies of factors affecting the occurrence, progression or outcome of disease.
9. Investigations of putative environmental hazards including industrial sources of pollution.
10. Investigations of disease clusters or clustering.
11. Selection of geographically ordered population samples for surveys.

A central feature of these analyses is the geographical linkage of health data with that on population characteristics, environmental conditions and health care. Such linkage is one of the principal capabilities of GIS which has considerably extended the scope and sophistication of these analyses. The range of health data used in them is wide and includes statutory registrations of deaths, cancer registrations, births and stillbirths, congenital malformations, hospital admissions, patient contacts in primary and community care (including data on screening and immunisations), notifications of communicable and sexually transmitted diseases, prescription data, and data from special surveys or surveillance programmes. One of the key developments of recent years in the United Kingdom that has contributed so much to our ability to analyse these data geographically has been the increasingly widespread use of the postcode as a spatial marker.

Various of these routine datasets (for example deaths, cancer registrations, births and hospital admissions) contain the individual's full unit postcode of residence which locates that individual to around 14 households or 40 individuals. With the availability of computerised directories giving the coordinates of postcode centroids accurate to 100 metres or less, this coding means that such data can effectively be treated as points. They can therefore be analysed as such, readily linked with other spatially-referenced data, and flexibly aggregated into larger geographical units as desired. Coupled with the technological advantages of GIS, this has provided the potential for a wide variety of new types of geographical analysis hitherto impracticable in the routine setting.

In the United Kingdom, population data used in geographical studies come from two main sources:

1. The census which provides data on the age and sex structure of the population, socioeconomic data (for example for calculating the under-priveleged area (UPA) score or Carstairs index, Carstairs (1993); Jolley et al. (1992)) and other demographic information such as ethnic composition.
2. The Family Health Services general practice register, containing the individual's postcode of residence as well as the UPA score for his/her ward of residence.

Of these two, the census is the more accurate and comprehensive and provides data down to the level of enumeration districts – approximately 170 households or 400 persons. However, its accuracy becomes increasingly uncertain towards the end of the inter-census period, it does not directly relate to general practice populations (which are geographically dispersed), and unlike the FHS register, it cannot provide a sampling frame for population surveys. Sampling frames can, however, be obtained from sources such as the electoral register and computerised files of address locations (for example Royal Mail Postcode Address File, Ordnance Survey ADDRESS-POINT data). Because of inherent inefficiencies in identifying transfers of patients between areas, the FHS register can be inaccurate and the number of patients on it may be considerably larger than that estimated from the census. This 'list inflation' can exceed 50% in some more mobile age groups, especially young adults. For most purposes the census remains the 'gold standard'.

13.2 Role of GIS

The potential applications of GIS in geographical studies of health are many. Some are described below. Discussion here concentrates on the advantages and disadvantages of using GIS in these, rather than on technical details of implementation. The aim is to provide a broad view of the potential of GIS in geographical studies of health.

13.2.1 Disease mapping and geographical correlation studies

One of the most useful functions of GIS in public health is its most basic – mapping. Maps of health statistics can be invaluable in understanding local patterns of disease and their geographical associations. They have the advantage of conveying instant visual information accessible to non-experts as well as public health professionals though their interpretation can be far from straightforward. Issues relating to data quality and sparse numbers are particularly important at the small area level; and the visual impact of mapped data is influenced by uneven population distribution which leads to disproportionate emphasis on large sparsely-populated areas and a greater likelihood of observing apparently extreme risks in such areas. The map layout, content and data representation also influence interpretation.

In many health applications, a crucial issue is data quality: the use of complete and reliable data based on standard disease definitions is pre-requisite. We still have only limited knowledge of the quality of most routine health data such as deaths and cancer registrations and the uncertainties are greater where there is relatively little quality control as with general practice data on chronic disease management.

We are now able to analyse data at a fine resolution. However, this magnifies any problems with data integrity, and the infrequency of most health events means that true variation in disease rates among small populations can be difficult to disentangle from random noise. The precision of a rate depends on the number of cases or events, and hence is generally poorer for a rare than a common disease. But even for quite common conditions, such as those included in the Health of the Nation strategy, numbers within a given area can be small. For example, in a locality of around 50 000 people, only five deaths from stroke under 65 years would be expected annually, only four cases of invasive cervical cancer, and three breast cancer deaths in the 50–64 age band (the group targeted in the UK national screening programme). Precision can, of course, be improved by aggregating over time or space. Methods of map smoothing are sometimes relevant in this context, and may be based on statistical techniques (for example empirical Bayesian smoothing; see Clayton and Kaldor (1987) and López-Abente, chapter 10) or GIS methods (for example local-area averaging). The scale of analysis or level of aggregation is a trade-off between specificity and precision: the smaller the area the more specific and relevant are the findings to the population that live within it, but the greater the imprecision and potential for bias.

Because numbers are often small in individual localities, studies which draw on data from many areas are generally more robust. A benefit of GIS is that it allows semi-automated data processing, so that analysis can be carried out at high resolution *and* with wide geographical coverage. Geographical correlations quantify the relationship between disease and its area-based determinants. An important application is the investigation of wealth-related disparities in health (Sloggett and Joshi, 1994) which have tended to widen over the last decade rather than diminish (Whitehead, 1988). However, in geographical correlation studies the measure of association between 'exposure' and disease at group (area) level may be a distortion of the association at individual level (Greenland and Morgenstern, 1989). This ecological bias occurs when the background rate of disease varies between populations because of differences in the distribution of other risk factors, or when the effects of a particular 'exposure' are modified by other factors.

Some of the interpretational issues are illustrated in Figure 13.1 which shows a ward-level analysis of hospital admissions with asthma, 1992–1993, in a west London health district (Landon, 1996). Hospital admission ratios have been standardised by five-year age group and sex. There is wide variation in the point estimates of standardised hospital admission ratios (SARs) for individual wards, ranging from 28 to 215, and wide scatter about the regression line. The scatter reflects the small annual number of asthma admissions in each ward but also data errors and systematic differences between hospitals in the recording of asthma. Part is also presumably related to variation in risk factors and health care. It is difficult to draw firm conclusions about the standardised admission ratio for an individual ward, but when mapped, certain patterns emerge which reflect the more general relationships, namely that hospital admission with asthma is more likely in wards of greater socioeconomic deprivation and large south Asian (New Commonwealth) populations. These relationships may be useful to health authorities in deciding how the need for health care varies from area to area (Brownson et al., 1992) and thus how resources should be most equitably distributed (Carr-Hill et al., 1994; Judge and Mays, 1994; Smith et al., 1994).

Interestingly, there appears to be no relationship, or even a negative one, between local road density (computed using GIS methods) and asthma admission. The problem for interpretation is that it is unclear whether the slope truly represents the association between air pollution and asthma, or whether it is biased by some unmeasured risk factor or variation in data quality. We will return to this point below. The main purpose of disease mapping is usually to gain insights into the spatial distribution of disease determinants, or to develop hypotheses that may be tested using other methods.

13.2.2 Patterns of health service use and access

In many ways, GIS is an ideal tool for studying organisational issues of the health service (Twigg, 1990; Hirschfield et al., 1995) though its full potential is not always exploited by health planners. Three example applications are described.

Referral patterns

Questions about patterns of hospital referral, clinic locations and many other issues germane to health service organisation are highly suited to GIS analysis. As an example, in cancer medicine there are questions about the optimal size, structure and location of treatment centres. There are good reasons to concentrate some services in the hands of a small number of specialists who see a reasonably high volume of cases and who are therefore able to develop the experience and facilities to offer high quality clinical care. In consequence, the best organisation of services within a district may entail some hospitals having a specialist unit for a particular cancer but not others. Different hospitals may develop specialisation in different areas of work. How services are most effectively organised – which hospitals should

GIS IN PUBLIC HEALTH 183

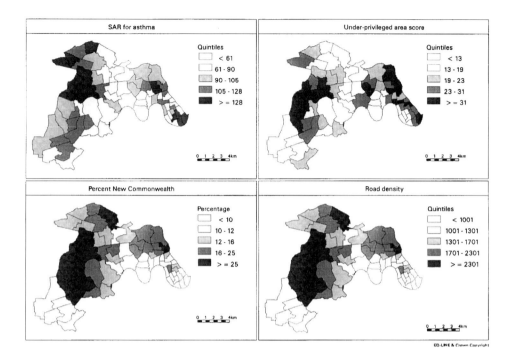

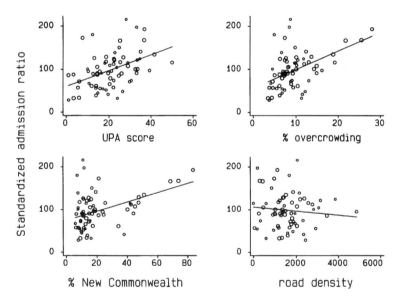

Figure 13.1

develop specialist facilities, where clinics should be sited etc. – depends on many factors, some of which are clearly geographical: how people currently use cancer services in different areas and how re-structuring them could affect their access to it. GIS may be used to examine current patterns of use and to carry out modelling of different 'scenarios', taking account of the population distribution, location of facilities, transport links, primary care services and a range of other factors.

Another example is the re-allocation of general practice patients to other practices when, for example, a single-handed practitioner retires. Again, useful guidance can be gained from exploring the geographical distribution of registrants in relation to alternative primary care facilities. GIS can be directly used to select and re-assign patients to specific practices.

Estimation of service areas

The population served by a hospital is usually geographically dispersed and not neatly constrained by administrative or census boundaries. There are thus problems of accurate enumeration. Nonetheless, it is often desirable to characterise the population that a hospital serves so that proper account can be taken of its needs. Overlay of map layers – patient address locations on census boundaries and other geographical features – using GIS techniques can help to define a hospital's 'service area'. The base population within that area may then be constructed by summing the populations of component areas (enumeration districts or wards). Various rules can be used, but a common principle is to select those areas in which a greater number of admissions go to the index hospital than to any other. Visual inspection of the distribution of individual admissions aids this process.

Service areas defined in this way allow comparisons of hospital admission rates with adjustment for the age, sex and socioeconomic structure of the population. It is worth noting, however, that boundaries of service areas differ for emergency and elective admissions, are disease and specialty-specific, and they frequently change over time. Indeed, changes in service areas may be of direct interest to health planners, particularly given the theoretical freedom of patients to choose where they go for treatment.

General practice populations are also geographically dispersed and, as with hospital admissions, it is possible to use the postcode of residence to define 'localities' in which the majority of residents are registered with a local general practitioner. Localities have been particularly relevant in the UK with the move towards the local commissioning of health services (Grundy et al., 1995; Kivell et al., 1990). The selection of locality boundaries is a complex process in which GIS (zoning) analysis can play an important role: it is desirable that localities reflect where people are treated; that they are relatively homogeneous in socio-demographic characteristics; that they adhere to natural boundaries (roads, waterways, etc.); that they have a natural sense of community and are relevant in terms of the need for health care.

Access

Two related concepts of access, relevant to health care organisation and analysable using GIS, are accessibility of services and equity in their provision. It is clearly desirable that primary, secondary and community health care services are accessible to all sections of the population including those who have to rely on public transport. Areas served by specific facilities and travel times to them are therefore important to service configuration, and can be examined in the framework of location analysis (Eyles, 1990; Armstrong et al., 1992). Social as well as physical factors can be brought into analysis using area-based markers of socioeconomic deprivation, single parent families, households without access to a car, etc.

A more general issue is the degree to which health care is evenly spread (equity of supply). GIS methods can help generate appropriate measures of access and define variation in it. Such measures are typically based on the supply (for example number of hospital beds) in relation to the population density, with appropriate weightings for distance. Measures of this kind were used in the development of a resource allocation formula for hospital in-patient services in the UK.

13.3 Studies of environmental hazards and disease clusters

The geographical investigation of environmental health hazards encompasses a wide range of studies in which GIS can play a useful role, including studies of reported disease 'clusters', geographical surveillance, analyses of health statistics in relation to single or multiple sources of pollution, geographical correlations and investigations of chemical accidents, as several chapters in this volume illustrate (see also Vine et al., 1997; Briggs and Elliott, 1995).

Cluster investigations, initiated in response to reports of apparent disease excess in a locality, are often demanded by public concern, but are difficult to interpret and seldom provide new insights into the determinants of disease (Neutra, 1990; Rothman, 1990). Their *post hoc* nature in general precludes a valid assessment of statistical inference, and the cluster itself can be critically dependent on the selection of boundaries in time and space. The significance of a local disease excess has to be judged more from the pattern of cases (their diagnostic specificity, spatial association with a potential hazard, the overall magnitude of the excess, etc.) than from statistical tests; and the findings should also be considered in the light of other epidemiological and toxicological evidence. Because clusters are intrinsically defined on the basis of aggregation in space and time, GIS may be helpful for data exploration, and integrating health data with other relevant datasets for formal epidemiological analysis.

Computer-driven search procedures based on GIS technology have been developed for scanning health statistics to determine the locations of unusual case aggregations (Openshaw, 1990). Though theoretically attractive as an early warning system or surveillance tool, issues relating to multiple testing, uncertainties of data quality and the limited capacity to adjust for known risk factors, lead to difficulties of interpretation. 'Hot spots' are almost bound to be uncovered, yet in most cases no cause will ever be identified. There is clearly a balance to be struck between generating unwarranted public concerns and identifying genuine health risks as early as possible. Further research will help to clarify what, if any role, such methods have; but the prevailing view among epidemiologists is against their use given the present state of scientific knowledge.

Researchers have also developed GIS-based systems as a support tool for geographical surveillance (for example, Rushton, chapter 5). Typically, such systems have capabilities for calculating disease rates in operator-determined areas, and for applying various statistical tests to them. Their use is likely to grow as the technology becomes more and more accessible, but the interpretation of their output again requires expert judgement and considerable circumspection.

Where concerns focus on the effects of specific pollution sources (usually industrial sites), GIS methods may be useful for developing markers of exposure (Briggs, 1992). Some examples in the field of air pollution include use of buffering zones to determine areas close to roads of high traffic density; production of a pollutant emissions grid based on a model of mobile and stationary emission sources; modelling of the dispersion of pollutants from a point source such as a smoke stack; and generation of pollution contours by interpolation of measured concentrations (Figure 13.2).

Ideally, data on the spatial distribution of pollution is integrated with data on the distribution and movement of individuals. Unfortunately, this is seldom possible. Problems of population movement are especially great where there is a long latency between exposure and disease onset. Lung cancer, for example, typically occurs two or three decades after exposure, and migration may result in substantial exposure misclassification if it is determined on the basis of current place of residence (see Löytönen, chapter 7).

Even with short-term effects, the movement of individuals through the environment may again mean that place of residence is a poor marker of integrated exposure. Many studies of environment and health rely on simple proximity of residence to the pollution source as a measure of relative exposure. The consequent misclassification is likely to lead to biased estimates of the relation between exposure and

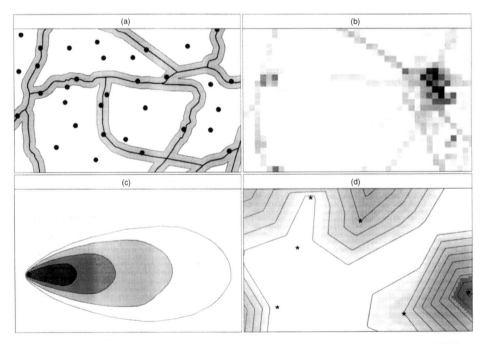

Figure 13.2 Examples of GIS use in exposure modelling: (a) buffer zones around roads of high traffic density (address locations shown as dots); (b) grid data from a pollutant emissions model; (c) pollutant dispersion from a point source; (d) pollution contours from interpolation of concentration measurements at fixed sites (stars).

disease. However, using GIS it is possible to generate a marker of exposure for very large numbers of individuals in different settings, and judicious selection of the scale of analysis may help to reduce misclassification: air pollution measurements in different cities, for example, are probably a reasonable reflection of the relative average exposure of people living within them.

Spatial analyses of disease risk are also often hampered by the limited availability of confounding factor data. Census data usually allow adjustment for socioeconomic deprivation, but reliable data on smoking prevalence are generally unavailable. On the whole, risks due to environmental exposures are very small, and this is an important consideration in the design of GIS-based epidemiological studies of environment and health.

13.4 Modelling health impacts

An application of GIS which so far has not been well-developed in public health but which is a subject of increasing interest is its use in modelling health impacts of environmental hazards. GIS is useful for modelling the distribution of exposures as well as the size and characteristics of the population at risk. It has application in areas such as the quantitative risk assessment of exposure to outdoor air pollution, the estimation of the health impact of flood hazards, and the health consequences of local transport policies. In research settings, workers have begun to explore the ecology of infectious diseases and effects of global climate change (Thomson et al., 1996; Randolph, 1993; Jetten et al., 1996) – for example the spread of vector-borne diseases, heat-related and cold-related deaths, changes in agricultural patterns and yields, and the effects of ultraviolet radiation from ozone depletion.

GIS methods are also being used in planning the response of emergency services to major chemical spillage and similar accidents. Modelling of different scenarios of pollution dispersion from an accident provides an indication of expected casualties and later health impacts. This information can then be used to test in theory the capacity of emergency services and health systems to deal with the emergency. Useful lessons may be learned from such case studies about emergency response procedures.

13.5 Conclusions

Geographical variation in health and the need for health care are influenced by a variety of social, environmental and health care factors. Geographical analysis of health variation and its determinants thus has an important role in public health; GIS can play a valuable role in this. Mapping of disease status, while one of the simplest functions of GIS, is an important aid in understanding local patterns of disease, though interpretation can be problematic. Issues of data quality may be particularly important, especially at the small area level.

The use of GIS in health service planning is expanding. Patterns of health care use can be examined and location analysis used to aid decisions about the optimal configuration of services, taking account of the population distribution, location of current health care facilities, transport links and other factors. It may also be helpful in analysing access to health services and equity in their provision.

GIS has many uses in environmental epidemiology, especially in generating markers of environmental exposure, though further developmental work is required in areas such as integrated exposure modelling and geographical surveillance. Work is expanding on the use of GIS to model health impacts of environmental hazards and public policies.

Data issues are important to future development. High-resolution geo-coding of routine health datasets has been one of the important contributors to the expansion of GIS use in health research in the UK and Scandinavian countries. It is to be hoped that similar accuracy of geographical referencing will in time become more widely established. Future development needs to focus on data quality, and the collection of data on a broader range of health endpoints, including those relating to quality of life and perception of illness. Further methodological work is also required on the integration of space and time data, particularly in relation to environmental exposure, and on the determinants of background variation in disease.

References

ARMSTRONG M. P., DENSHAM P. J., LOLONIS P. and RUSHTON G. (1992) Cartographic displays to support locational decision-making. *Cartographic and Geographic Information Systems*, **19**, 154–164.

BRIGGS D. J. (1992) Mapping environmental exposure. Pages 158–176 in Elliott P., Cuzick J., English D. and Stern R. (eds) *Geographical and Environmental Epidemiology: Methods for Small-Area Studies*, Oxford University Press, Oxford.

BRIGGS D. J. and ELLIOTT P. (1995) The use of geographical information systems in studies on environment and health. *World Health Statistics Quarterly*, **48**, 85–94.

BROWNSON R. C., SMITH C. A., JORGE N. E., DEPRIMA L. T., DEAN C. G. and CATES, R. W. (1992) The role of data-driven planning and coalition development in preventing cardiovascular disease. *Public Health Reports*, **107**, 32–37.

CARR-HILL R. A., SHELDON T. A., SMITH P., MARTIN S., PEACOCK S. and HARDMAN G. (1994) Allocating resources to health authorities: development of method for small area analysis of inpatient services. *British Medical Journal*, **309**, 1046–1049.

CARSTAIRS V. (1993) Deprivation indices: their interpretation and use in relation to health. *Journal of Epidemiology and Community Health*, **49**, S3–S8.

CLAYTON D. and KALDOR J. (1987) Empirical Bayes estimates of age-standardized relative risks for use in disease mapping. *Biometrics*, **43**, 671–681.

EYLES J. (1990) How significant are the spatial configurations of health care systems? *Social Science and Medicine*, **30**, 157–164.

GREENLAND S. and MORGENSTERN H. (1989) Ecological bias, confounding and effect modification. *International Journal of Epidemiology*, **18**, 269–274.

GRUNDY C., LANDON M., THAKRAR B., CRAWLEY R. and WILKINSON P. (1995) The use of geographical information systems for assessing health care needs at locality level. *Proceedings of the Association for Geographic Information*, 3.2.1–3.2.3.

HIRSCHFIELD A., BROWN P. J. B. and BUNDRED P. (1995) The spatial analysis of community health services on the Wirral using Geographic Information Systems. *Journal of the Operational Research Society*, **46**, 14–59.

JETTEN J. H., MARETNS W. J. M. and TAKKEN W. (1996) Model simulations to estimate malaria risk under climate change. *Journal of Medical Entomology*, **33**, 361–371.

JOLLEY D. J., JARMAN B. and ELLIOTT P. (1992) Socio-economic confounding. Pages 115–124 in Elliott P., Cuzick J., English D. and Stern R. (eds) *Geographical and Environmental Epidemiology: Methods for Small-Area Studies*, Oxford University Press, Oxford.

JUDGE K. and MAYS N. (1994) Allocating resources for health and social care in England. *British Medical Journal*, **308**, 1363–1366.

KIVELL P. T., TURTON B. J. and DAWSON B. R. P. (1990) Neighbourhoods for health service administration. *Social Science and Medicine*, **30**, 701–711.

LANDON M. (1996) Intra-urban health differentials in London – urban health indicators and policy implications. *Environment and Urbanization*, **8**, 119–128.

NEUTRA R. R. (1990) Counterpoint from a cluster buster. *American Journal of Epidemiology*, **132**, 1–8.

OPENSHAW S. (1990) Automating the search for cancer clusters: a review of problems, progress and opportunities. Pages 48–79 in Thomas R. (ed.) *Spatial Epidemiology*, Pion, London.

RANDOLPH S. E. (1993) Climate, satellite imagery and the seasonal abundance of the tick *Rhipecephalus appendiculatus* in southern Africa: a new perspective. *Medical and Veterinary Entomology*, **7**, 243–258.

ROTHMAN K. J. (1990) A sobering thought from a cluster buster. *American Journal of Epidemiology*, **132**, S6–S13.

SLOGGETT A. and JOSHI H. (1994) Higher mortality in deprived areas: community or personal disadvantage? *British Medical Journal*, **309**, 1470–1474.

SMITH P., SELDON T. A., CARR-HILL R. A., MARTIN S., PEACOCK S. and HARDMAN G. (1994) Allocating resources to health authorities: results and policy implications of small area analyses of inpatient services. *British Medical Journal*, **309**, 1050–1054.

THOMSON M. C., CONNOR S. J. and MILLIGAN P. J. M. (1996) The ecology of malaria – as seen from earth observation satellites. *Annals of Tropical Medicine and Parasitology*, **90**, 243–264.

TWIGG L. (1990) Health-based geographical information systems: their existing potential examined in the light of existing data sources. *Social Science and Medicine*, **30**, 143–155.

VINE M. F., DEGNAN D. and HANCHETTE C. (1997) Geographic Information Systems: their use in environmental epidemiologic research. *Environmental Health Perspectives*, **105**, 598–605.

WHITEHEAD M. (1988) The health divide. In *Inequalities in Health*, Penguin, London.

CHAPTER FOURTEEN

Improving Health Needs Assessment using Patient Register Information in a GIS

ANDREW LOVETT, ROBIN HAYNES, GRAHAM BENTHAM, SUE GALE, JULII BRAINARD and GISELA SUENNENBERG

14.1 Introduction

Much research on environmental epidemiology and public health requires geographically detailed and up-to-date information on population characteristics. Across Europe the manner in which such data are collected and made available varies considerably (Redfern, 1989; Waters, 1995). Some countries use decennial censuses, while others rely on registers or a mixture of the two methods (see Table 14.1). Organisations such as Eurostat are also involved in projects to compile demographic statistics for areas across Europe, but ensuring reasonable consistency in these databases is not straightforward and the geographical resolution is unlikely to extend below the NUTS5 (community) level.

In the United Kingdom (UK) a decennial census (last taken in April 1991) has been the traditional source of population information (Dale and Marsh, 1993; Openshaw, 1995). This has the advantage of covering a substantial number of topics and providing statistics for areas down to the size of an Enumeration District (on average around 400 persons). Even such areas, however, can be larger than is ideal for some epidemiological purposes (for example estimating at-risk populations around a point source in a rural area). A further complication in England and Wales is that the census zones do not match those of the postcode system which is widely used to georeference health records (Mohan, 1993). This difficulty, however, is now less pronounced due to the availability of directories which list a grid reference and best matching ED for each postcode (Martin, 1992), and the ability to undertake point-in-polygon and other intergrative operations within a GIS (Gatrell *et al.*, 1991; Gatrell and Dunn, 1995).

A second constraint with census data is the manner in which the information inevitably becomes less useful with the passage of time. Estimates of demographic

Table 14.1 Availability of official demographic data in western European countries

Country	Date	Source	Area name	Number of areas	Av. pop. (000)
Austria	1991	Census	Gemeinde	2333	3
Belgium	1991	Census	Communes	596	17
Denmark	1991	Register	Kommuner	276	19
Finland	1991	Register	Municipalities	445	11
France	1990	Census	Communes	36 545	2
Germany	1991	Surveys	Gemeinde	16 147	5
Greece	1991	Census	Demoi/Koinotites	6039	2
Ireland	1991	Census	DED Wards	3438	1
Italy	1991	Census	Comuni	8097	7
Luxembourg	1991	Census	Commune	118	3
Netherlands	1991	Register and Survey	Gemeenten	702	21
Norway	1991	Census and Register	Municipalities	440	10
Portugal	1991	Census	Concelhos/Municipios	305	34
Spain	1991	Census	Municipios	8056	5
Sweden	1990	Census and Register	Kommuns	284	30
Switzerland	1991	Census	Communes	3021	2
UK	1991	Census	Wards	10 970	5

Source: based on Waters, 1995, p. 41.

structure (for example totals for age groups) can be made by projecting forward population cohorts, but inevitably there are limits (particularly at the local level) in the extent to which migration can be taken into account (OPCS, 1991). In the UK, the Office of National Statistics does not publish mid-year population estimates below the scale of local authority districts (for example ONS, 1996), though various public and commercial organisations have produced ward estimates for particular cities or regions on an *ad hoc* basis (Woodhead and Dugmore, 1990; Birkin et al., 1996). On average a ward contains around 5000 people, but these areas are often socially heterogeneous and are certainly larger than desirable for some health research purposes (Reading et al., 1990).

Another problem can be refusal to cooperate with censuses (Redfern, 1989). The 1991 census in the UK is thought to have undercounted the population by about 2% (one million people), perhaps because of the widespread but mistaken belief that census questionnaires would be used to help compile lists of those required to pay a new per capita local government tax. This undercoverage also varied with age, sex and geographical area, the worst case being that of males aged 25–29 in London and other major conurbations where it is estimated that only about 80% of those present were included in the census (OPCS, 1994).

Due to these issues, there has been some interest in the potential of administrative registers as a source of UK population data (OPCS, 1993). The most relevant database is the National Health Service Central Register which consists of details provided by Family Health Service Authorities (FHSAs) regarding people registered with general practioners (GPs). In theory all persons should be registered with a GP. There are, however, doubts about the reliability of this information. For instance, inflation of patient registers due to the administrative delay in removing outdated records is recognised as a source of inaccuracy (Fraser and Clayton, 1981;

Sheldon et al., 1984). The problem is likely to be worst in inner city areas with high population turnover (Bowling and Jacobson, 1989; Bickler and Sutton, 1993) where in addition considerable numbers of people may not be registered with any general practitioner (London Health Planning Consortium Study Group, 1981). Unfortunately, the group of people least likely to register with a doctor (young men in urban areas) are also those whose inclusion in the most recent census is most doubtful. Until recently, another limitation was incomplete postcoding of patient addresses in many registers. During the 1990s, however, changes in the organisation of health services have increased the need for accurate data on local residents and this, in turn, has stimulated efforts to improve the quality of patient registers (Wrigley, 1991).

Given this context, recent research at the University of East Anglia has sought to assess the reliability of the FHSA registers in a rural region and then evaluate the scope for using such data to improve the derivation of health needs indicators. Much of this work would not have been possible without the integrative capabilities of a GIS and it illustrates the very practical contribution such a tool can make to public health medicine. The next section of this chapter therefore discusses how population estimates were produced from the FHSA registers. This is followed by two examples of need indicator generation and the chapter concludes with some assessment of the prospects for wider use of this type of demographic data.

14.2 Deriving population estimates from patient registers

An assessment of register coverage for the counties of Norfolk and Suffolk in eastern England was made through a comparison with results from the 1991 census. Copies of part of the patient registers current on census day (21 April 1991) were obtained from Norfolk and Suffolk FHSAs. The records were anonymous, consisting of each patient's date of birth, sex, postcode of residence and the code number of the patient's GP. Where possible, each record was assigned to a census ward on the basis of its postcode. One method of matching postcodes to census wards is to use the Central Postcode Directory (Raper et al., 1992) but this is known to produce inaccuracies (Gatrell, 1989). We therefore compared the results from this method with three other approaches. These were: use of the Postcode Address File (Raper et al., 1992), use of the first release of the OPCS Postcode-Enumeration District Directory (Martin, 1992), and use of a point-in-polygon method to locate the national grid coordinates of each unit postcode (with the addition of 50 m to both northings and eastings to improve accuracy) within digitised ward boundaries by means of a geographic information system (Gatrell et al., 1991). For 93% of postcoded records in Norfolk and 91% in Suffolk the four methods produced the same ward allocation.

Each of the remaining postcodes was assigned to the wards on which at least two independent methods agreed. Preference was given to data with different antecedents so that, for example, agreement between the point-in-polygon method and the OPCS directory was regarded as more significant than that between the Central Postcode Directory and the Postcode Address File. In many situations where there were three or more possible ward allocations for a postcode it was concluded that a tenable choice between the alternatives could not be made. Problems also existed because some postcodes (involving fewer than 1% of postcoded patients) could not be traced in any database and others were placed by the Post Office in different counties to those anticipated. Such records with postcodes which could not confidently be assigned to a ward, together with records with no postcode, were allocated to wards in proportion to the already assigned population registered with the same GP practice and in the same age and sex category. More patients were allocated by this method in Norfolk than Suffolk (see Table 14.2) and the final totals were slightly different from the original patient numbers due to rounding in the calculations (Haynes et al., 1993).

The 1991 census population counts for Norfolk and Suffolk wards were obtained from the Local Base Statistics files. The counts were those for usually resident population: 1991 base. To take account of enumerated residents not expected to appear in FHSA patient registers, the usually resident population was adjusted by subtracting resident students with a term time address outside the area, adding students resident outside the area but living in the area during term time, and subtracting hospital patients, prison residents and those employed in the armed forces (Haynes et al., 1995a). The census totals after adjustment are shown in Table 14.2 and indicate that both registers contained more patients than were enumerated in the 1991 census.

Table 14.3 shows that the ratio of census to register totals varied between age groups. Census totals were markedly higher than register totals for males and females aged 0–4 years, and they were lower for young adults aged 15–44 years (especially men) and for males and females aged over 74 years. For other age groups the match was close.

The higher census populations (compared with patient register totals) for children aged 0–4 years can be explained by delays in entering births onto patient registers. Similarly, larger numbers of registered patients aged over 74 years may well reflect lags in removing recently deceased people from the register. Differences in the totals for young adults (particularly men aged 15–44 years) were probably a consequence of census undercounting outweighing any tendency for this group not to register with a doctor. At the overall county scale, then, the patient register esti-

Table 14.2 Comparison of population totals

County	Number of patients	% Patients unpostcoded	Final estimate	Adjusted census
Norfolk	757 042	7.9	756 857	736 948
Suffolk	636 642	1.9	636 645	622 015

Table 14.3 Census population as a % of register estimate

	Norfolk		Suffolk	
Age	Males	Females	Males	Females
0–4 years	104	105	106	106
5–14 years	99	98	100	101
15–44 years	93	97	93	98
45–64 years	99	100	97	99
65–74 years	99	99	98	99
Over 74 years	97	96	96	96

Table 14.4 Cumulative percentage of register ward population estimates within specified range of the census count

	Percentage within		
	5%	10%	20%
Norfolk	53	85	95
Suffolk	59	83	95

mates of the population appear equivalent or even superior to census counts for most age groups, except for children aged 0–4 years (where register totals need to be increased) and for elderly people aged over 74 years (where register totals should be reduced).

Differences between census counts and patient register totals were also examined at the ward scale. In the 1991 census there were 230 wards in Norfolk and 191 wards in Suffolk. It was hypothesised that census counts would be larger than register estimates in wards with high rates of in-migration (due to a lag in patient registrations) and those with substantial proportions of people employed in the armed forces (because of the availability of primary care for some dependents from

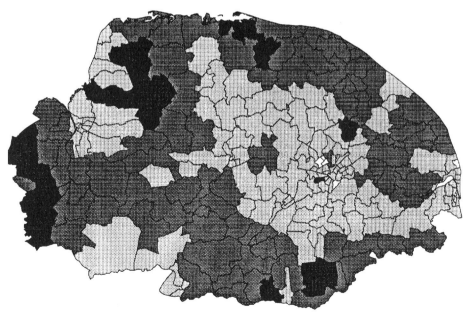

Figure 14.1 The contribution of the procedure for allocating unpostcoded patients to ward population estimates in Norfolk.

the armed forces' medical service). Both these trends were confirmed by regression analyses (Haynes et al., 1995a). As anticipated from research on underenumeration (OPCS, 1994), relatively low census counts compared with register totals were consistently associated with wards with high unemployment rates, large proportions of households in shared dwellings and high population density. Once these three sources of bias were taken into account, the remaining differences between census and register ward population estimates were distributed as shown in Table 14.4.

Only just over half of the wards had population register totals within 5% of the census count, but 95% were within 20% of the census total. Further examination revealed that many of the most extreme differences occurred in wards with either high proportions of armed services personnel or where the allocation method (for patients with unit traceable postcodes) was of particular importance. In 1991, the completeness of postcoding tended to vary between GP practices and so, as Figure 14.1 illustrates for the wards of Norfolk, the influence of the allocation procedure on the patient register population estimates displayed distinct positive spatial autocorrelation. Figure 14.2 shows the geographical pattern of patient register differences from the census totals for the same set of areas and provides several examples (for example at Blakeney on the northern coast) where substantial discrepancies coincide with higher values on Figure 14.1. Beyond the military base and allocation pro-

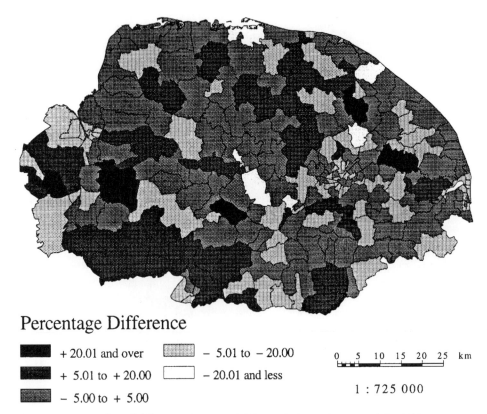

Figure 14.2 Unexplained differences between FHSA population estimates and the 1991 census as a percentage of the census population.

cedure influences, however, the variations in Figure 14.2 are not readily amenable to further interpretation.

Taken together, these results indicate some of the difficulties in using FHSA patient registers to produce small-area population estimates. The degree of agreement, however, is stronger than might have been anticipated from some previous comments on the reliability of patient registers and it is arguable that they are becoming ever more viable as a source of demographic information. Comparisons with census data serve as a check on the reliability of such databases and, moreover, they have the advantage that individually postcoded records can be aggregated with great geographical flexibility. Patient registers are also continuously updated and with the improvements that have occurred since 1991 (for example the Norfolk register is now over 99% postcoded so eliminating the allocation problem noted above) it is quite possible that in (at least) some parts of the UK they are now the best source of current basic demographic data. Of course, the census provides details of a far wider range of socioeconomic variables, but utilisation of patient register information can also contribute to the production of better needs indicators. Two brief illustrations of this point are given in the following sections.

14.3 Unemployment rate as an updatable social indicator

Many reviews of health service provision and resource allocation in the UK have used composite deprivation indices derived from census variables as need indicators (Mohan, 1993). Several studies, however, have suggested that unemployment rates are at least as appropriate as the composite measures (for example Campbell *et al.*, 1991; Payne *et al.*, 1993), but figures based on the International Labour Organisation (ILO) definition of unemployment are only available at a fine geographical scale (for example wards) through the decennial census. Such unemployment rates for small areas may quickly become outdated, but there is the possibility of a more up-to-date approximation because counts of people claiming state unemployment benefit are regularly published down to the ward scale. Unfortunately, eligibility to claim state unemployment benefit has been subject to many changes by the UK government in recent years (Bartholomew *et al.*, 1995). It is well known that people eligible for benefit are not necessarily unemployed according to the ILO definition of unemployment and, conversely, that only a proportion of those who fit the international criteria are able to claim unemployment benefit in the UK. The match is particularly poor for women (Woolford and Denman, 1993). There is consequently some doubt whether benefit claimants are as suitable a proxy for deprivation. The calculation of updatable unemployment rates based on claimant statistics also requires a compatible population denominator. An evaluation of these issues was undertaken using data for wards in East Anglia (Haynes *et al.*, 1996). At the time of the study (1994) unemployment claimant counts were only available for wards defined in terms of 1981 boundaries and so it was only feasible to assess trends in the 293 (of 576) wards whose boundaries did not change between 1981 and 1991. This subset was biased towards the more deprived urban areas, but this characteristic was not thought to invalidate the investigation. For each ward, male unemployment rates were calculated in the following three ways.

1. Unemployment as a percentage of the economically active population aged 16 or over (with both variables from the 1991 census).

2. Unemployment benefit claimants in April 1991 as a percentage of the male population aged 15–64 from the OPCS mid-year population estimates for 1991.
3. Unemployment benefit claimants in April 1991 as a percentage of the male population aged 15–64 estimated from FHSA patient registers.

Table 14.5 Correlations of unemployment rates and deprivation indices with health measures

	% unemp. Census	% unemp. FHSA	% unemp. MYE	Townsend	Carstairs	Jarman
SIR < 75 m	0.64	0.61	0.62	0.57	0.63	0.52
SIR < 75 f	0.66	0.64	0.63	0.67	0.69	0.66
SIR < 75 b	0.68	0.65	0.65	0.64	0.69	0.61
SMR < 75 m	0.48	0.43	0.46	0.40	0.47	0.39
SMR < 75 f	0.33	0.34	0.33	0.30	0.30	0.31
SMR < 75 b	0.53	0.50	0.53	0.47	0.51	0.45

Note: m = males; f = females; b = both sexes.

These measures, along with three composite deprivation indices, were then correlated with standardised long-term illness (SIR) and mortality (SMR) rates and the results are shown in Table 14.5.

Scrutiny of the correlation coefficients reveals that there is little to choose between the six indicators as predictors of health status. This supports the view that claimant rates may be as suitable a measure as ILO defined unemployment variables and more complex composite indicators. It also suggests that the male population aged 15–64 estimated from the patient register is a reasonable alternative unemployment denominator to the number of economically active men aged 16 years and older from the census, a point which is important since the former is much more readily updateable than the latter. These results were obtained using data for a census year (1991), and as such information becomes out-of-date in an inter-censal period the advantages of using current unemployment claimant rates (based on claimant numbers divided by population estimates from patient registers) as an indicator of health needs seem likely to become stronger still.

14.4 Calculating census health indicators for GP practices

Despite the possibilities provided by administrative registers, there are still many situations in the UK where the only practical source for a particular variable is the decennial census. In a health service context, however, it is often desirable to derive these variables for entities such as GP practices which serve overlapping populations dispersed across a substantial number of census areas (Scrivener and Lloyd, 1995; Birkin et al., 1996). Validated methods of translating census data into practice (or other service centre) estimates are therefore needed and patient registers can be of considerable value in such work.

Several options were examined using the April 1991 patient register data for 81 general practices in Suffolk (Haynes et al., 1995b). As a simple test, the percentage of patients aged 75 years and over was determined from the register data for each practice and these values were then compared with four different estimates derived from census data.

The first estimation method was simply to use the value of the ward in which the main surgery was located. The second was to calculate census values for practice areas, which have boundaries marking the normal geographical extent of the population served by each practice, as agreed between the doctors and the administrative authority (that is the FHSA). Practice areas typically contain the majority of, although not all, registered patients, but also include patients registered with other practices. Using definitions of practice areas supplied by the Suffolk FHSA, we calculated the percentage population aged 75 years and over for all wards within each practice area. The third and fourth methods involved using postcode addresses from the patient register to calculate the distribution of practice patients in wards and enumeration districts respectively. A look-up table relating Suffolk postcodes to wards and enumeration districts was created from Release 2 of the Office of Population Censuses and Surveys postcode-enumeration district directory. This was then linked to the Suffolk FHSA patient register of April 1991. The register contained 636 642 patient records and 618 455 were successfully matched with wards and 617 345 with enumeration districts. For each practice the numbers of patients in each ward and enumeration district were computed. Estimates of the percentage of patients aged 75 years and over were derived by multiplying each ward percentage population aged 75 and over by the number of patients in that ward, adding the products, and dividing by the total number of patients (third method). This procedure therefore sought to take account of the uneven distribution of patients between wards by weighting the census values by patient numbers. Similar calculations were performed using census data and patient numbers for enumeration districts (fourth method). On average, each practice had patients resident in 27 wards and 140 enumeration districts.

Subtracting actual register values from estimates for each practice, the limits of agreement (Bland and Altman, 1986) within which 95% of the differences were expected to lie were calculated (see Table 14.6). The surgery location method produced estimates of which 95% were expected to lie between -3% and 6% of the patient register values. Using practice areas improved the 95% limits of agreement to within 3%, while those based on the distribution of patients in wards and enumeration districts produced the most accurate estimates, generally within $+2\%$ of the register values. Measurement error thus became smaller as more sophisticated methods were used, but the enumeration district data did not produce substantially better estimates than those for wards. Other studies in London and Bradford, however, have recently suggested that calculations at the enumeration district scale

Table 14.6 95% limits of agreement between values of percentage population aged 75 years and over by practice (census estimates minus patient register values)

Census estimate	Lower limit %	Upper limit %
Surgery location	−3.2	6.3
Practice area	−3.3	2.8
Patient-weighted ward	−2.3	2.2
Patient-weighted ED	−2.0	2.2

do produce noticeably more accurate results (Majeed et al., 1995; Scrivener and Lloyd, 1995). This could reflect contrasts between these cities and the more rural nature of Suffolk, particularly greater social and demographic diversity within urban wards.

In reality it would never be necessary to estimate the ages of patients registered with a GP practice from the census, as they are known already, but this example does illustrate several methods that could be applied to census variables which are not directly available for practices and allows their accuracy to be assessed. It must be recognised, nevertheless, that even the best of the procedures discussed above have their limitations (Carr-Hill and Rice, 1995; Scrivener and Lloyd, 1995). One is the implicit assumption that the patients of each practice are representative of the populations in the areas where they live, something which will not hold if the choice of GP tends to vary according to the age, social status, ethnicity, etc. of residents. Another problem with some census variables, especially at the enumeration district scale, is that rates can be based on very small numbers and so barnardisation procedures (the quasi-random addition of -1, 0, or $+1$ to counts as a confidentiality measure) may introduce bias into the results. Even so, it seems clear that weighting procedures based on patient distributions are appreciably better than the simple use of surgery locations or practice areas and, in the absence of any straightforward alternative, they do appear capable of providing serviceable estimates of census characteristics for dispersed and overlapping populations.

14.5 Conclusions

This chapter has focused on how FHSA patient registers can be used to improve several aspects of health needs assessment. We have shown how patient registers can be used within a GIS to make population estimates, to help produce updateable health needs indicators, and to manipulate census information into descriptive measures for geographically dispersed groupings of patients. The subject is not a particularly glamorous one, but the provision of such information is fundamental to many applications of GIS in epidemiology and service planning. Up to now, most health research in the UK has relied upon the census for basic demographic and socioeconomic information, but the results summarised in this chapter suggest that patient registers may be a more viable alternative (or at least a valuable supplement) than has been previously thought. Of course, the quality of patient registers in other parts of the UK (notably large conurbations) may be poorer than those examined in this paper. Unfortunately, information on the status of other registers is piecemeal and often predates the considerable improvements which have occurred in the 1990s. A comprehensive assessment of the current accuracy of patient registers in different types of areas (for example rural, suburban, inner urban) would therefore be of considerable benefit in establishing whether such information would be suitable for use on a wider scale.

We believe that patient registers deserve increasing attention as a source of basic population data for small area epidemiological studies and health services research. The quality of such registers appears to be steadily improving and their up-to-date nature is a particular benefit in the years between censuses. It may also become increasingly feasible to enhance the analysis of health events by matching together records for the same individual from registers established for different purposes. An assessment of the technical problems involved in such linkage is currently being undertaken at the University of East Anglia in the context of research into accidents among young children in Norwich. Further research into the practicalities of using other registers in a similar manner would appear worthwhile. Ultimately though, it must be recognised that technical problems of record linkage or indeed the accuracy of patient registers are less a constraint on their wider use than public concerns regarding threats to freedom and privacy (Redfern, 1989). It may be possible to ease such worries through data protection measures or legislation on the freedom of information, but until there is public and political agreement in the UK on an expanded role for patient registers it seems inevitable that their use will be confined to studies undertaken either within or on behalf of health authorities. For the foreseeable future, therefore, analyses utilising such data are likely to be localised and *ad hoc* but, as this chapter has demonstrated, the enormous potential of patient registers as a data source for health research should not be in doubt.

Acknowledgments

Research discussed in this chapter was funded by a consortium of East Anglian health authorities and the Economic & Social Research Council (grant R000233158). For 1991 census data, we acknowledge the source: The 1991 Census, Crown Copyright, ESRC Purchase. We are also grateful to staff of the ESRC Data Archive and the Census Dissemination Unit at the University of Manchester Computer Centre for their help.

References

BARTHOLOMEW D., MOORE P., SMITH F. and ALLIN P. (1995) *Report of the Working Party on the Measurement of Unemployment in the UK*. London, Royal Statistical Society.

BICKLER G. and SUTTON S. (1993) Inaccuracy of FHSA registers: help from electoral registers. *British Medical Journal*, **306**, 1167.

BIRKIN M., CLARKE G., CLARKE M. and WILSON A. G. (1996) *Intelligent GIS: Location Decisions and Strategic Planning*. GeoInformation International, Cambridge.

BLAND J. M. and ALTMAN D. G. (1986) Statistical methods for assessing agreement between two methods of clinical measurement. *Lancet*, **i**, 307–310.

BOWLING A. and JACOBSON B. (1989) Screening: the inadequacy of population registers. *British Medical Journal*, **298**, 545–546.

CAMPBELL D., RADFORD J. M. C. and BURTON P. (1991) Unemployment rates: an alternative to the Jarman Index? *British Medical Journal*, **303**, 750–755.

CARR-HILL R. and RICE N. (1995) Is enumeration district level an improvement on ward level analysis in studies of deprivation and health? *Journal of Epidemiology and Community Health*, **49**, Suppl. 2, S28–S29.

DALE A. and MARSH C. (eds) (1993) *The 1991 Census User's Guide*. HMSO, London.

FRASER R. C. and CLAYTON D. G. (1981) The accuracy of age-sex registers, practice medical records and family practitioner committee registers. *Journal of the Royal College of General Practitioners*, **31**, 410–419.

GATRELL A. C. (1989) On the spatial representation and accuracy of address-based data in the United Kingdom. *International Journal of Geographical Information Systems*, **3**, 335–348.

GATRELL A. C. and DUNN C. E. (1995) Geographical information systems and spatial epidemiology: modelling the possible association between cancer of the larynx and incineration in north-west England. Pages 215–235 in de Lepper M. J. C., Scholten H. J. and Stern R. M. (eds) *The Added Value of Geographical Information Systems in Public and Environmental Health*, Kluwer Academic Publishers, Dordrecht.

GATRELL A. C., DUNN C. E. and BOYLE P. J. (1991) The relative utility of the Central Postcode Directory and Pinpoint Address Code in applications of geographical information systems. *Environment and Planning A*, **23**, 1447–1458.

HAYNES R. M., GALE S. H., LOVETT A. A. and BENTHAM C. G. (1996) Unemployment rate as an updatable health needs indicator for small areas. *Journal of Public Health Medicine*, **18**, 27–32.

HAYNES R. M., LOVETT A. A., BENTHAM C. G., BRAINARD J. S. and GALE S. H. (1995a) Comparison of ward population estimates from FHSA patient registers with the 1991 census. *Environment and Planning A*, **27**, 1849–1858.

HAYNES R. M., LOVETT A. A., BRAINARD J. S., HINTON J. C., BENTHAM C. G. and GALE S. H. (1993) *Derivation of Ward Population Estimates from Cambridgeshire, Norfolk and Suffolk FHSA Patient Registers and Comparison with the 1991 Census*. Health and Environment Research Group Report 4, University of East Anglia, Norwich.

HAYNES R. M., LOVETT A. A., GALE S. H., BRAINARD J. S. and BENTHAM C. G. (1995b) Evaluation of methods for calculating census health indicators for GP practices. *Public Health*, **109**, 369–374.

LONDON HEALTH PLANNING CONSORTIUM STUDY GROUP (1981) *Primary Health Care in Inner London*. London Health Planning Consortium, London.

MAJEED F. A., COOK D. G., POLONIECKI J., GRIFFITHS J. and STONES C. (1995) Sociodemographic variables for general practices: use of census data. *British Medical Journal*, **310**, 1373–1374.

MARTIN D. (1992) Postcodes and the 1991 Census of Population: issues, problems and prospects. *Transactions Institute of British Geographers*, **14**, 90–97.

MOHAN J. (1993) Healthy indications? Applications of census data in health care planning. Pages 136–149 in Champion A. G. (ed.) *Population Matters: The Local Dimension*, Paul Chapman Publishing, London.

OFFICE OF NATIONAL STATISTICS (1996) *Key Population and Vital Statistics 1993: Local and Health Authority Areas*. Series VS, No. 20, PP1 No. 16, HMSO, London.

OFFICE OF POPULATION CENSUSES AND SURVEYS (1991) *Making a Population Estimate in England and Wales*. OPCS Occasional Paper 37, OPCS, London.

OFFICE OF POPULATION CENSUSES AND SURVEYS (1993) *Report on Review of Statistical Information on Population and Housing (1996-2016)*. OPCS Occasional Paper 40, OPCS, London.

OFFICE OF POPULATION CENSUSES AND SURVEYS (1994) *Undercoverage in Great Britain*. 1991 Census User Guide 58, OPCS, London.

OPENSHAW S. (ed.) (1995) *Census Users' Handbook*. GeoInformation International, Cambridge.

PAYNE J. N., COY J., MILNER P. C. and PATTERSON S. (1993) Are deprivation indicators a proxy for morbidity? A comparison of the prevalence of arthritis, depression, dyspepsia, obesity and respiratory symptoms with unemployment rates and Jarman scores. *Journal of Public Health Medicine*, **15**, 161–170.

RAPER J., RHIND D. and SHEPHERD J. (1992) *Postcodes: The New Geography*. Longman, London.

READING R. F., OPENSHAW S. and JARVIS S. N. (1990) Measuring child health inequalities using aggregations of enumeration districts. *Journal of Public Health Medicine*, **12**, 160–167.

REDFERN P. (1989) Population registers: some administrative and statistical pros and cons. *Journal of the Royal Statistical Society: Series A*, **152**, 1–41.

SCRIVENER G. and LLOYD D. C. E. F. (1995) Allocating census data to general practice populations: implications for study of prescribing variation at practice level. *British Medical Journal*, **311**, 163–165.

SHELDON M. G., RECTOR A. L. and BARNES P. A. (1984) The accuracy of age-sex registers in general practice. *Journal of the Royal College of General Practitioners*, **34**, 269–271.

WATERS R. (1995) Data sources and their availability for business users across Europe. Pages 33–47 in Longley, P. and Clarke, G. (eds) *GIS for Business and Service Planning*, GeoInformation International, Cambridge.

WOODHEAD K. and DUGMORE K. (1990) Local and small area projections. Pages 65–75 in *Population Projections: Trends, Methods and Uses*, OPCS Occasional Paper No. 38, OPCS, London.

WOOLFORD C. and DENMAN J. (1993) Measures of unemployment: the claimant count and the LFS compared. *Employment Gazette*, **102**, 455–463.

WRIGLEY N. (1991) Market-based systems of health care provision, the NHS Bill, and geographical information systems. *Environment and Planning A*, **23**, 5–8.

CHAPTER FIFTEEN

Conclusions

ANTHONY C. GATRELL and MARKKU LÖYTÖNEN

We hope that the papers in this volume give something of the flavour of contemporary, mostly European, work on GIS in health research. There are at least three ways in which European collaboration might develop further in the area of GIS and health. The first, and most obvious, is to set up comparative studies of particular health problems, perhaps reviewing links in different settings between health and the physical environment, or between health outcomes and social environments. Can we put to work the same kinds of GIS-related methodologies to explore the same health problems, but in different European countries? Some work along these lines is on-going (for example, van der Veen et al., 1994), but there is scope for further collaborative research projects.

Related to this are cross-border small area studies. What are the possibilities of, and problems in, such studies? Is it worth exploring these in the context of a quite specific research question; for example concerning clusters of child or specific adult cancers? Clearly, such a study raises issues about the comparability of diagnostic behaviour and of disease registers, as well as the differences between the basic spatial units from which denominators might be sought. Nonetheless, given cross-boundary transfers of polluted air and water there may be scope for setting up local collaborations.

Third, collaboration may be fruitful in the development of methods of analysis, and the sharing of methodological expertise. To some extent this is already happening, notably in the use of common methods of small-area estimation considered in this volume by Braga and López-Abente. But we need much more open exchange of experiences of using particular methods; we also need much greater recognition, both in Europe and North America, as well as in other parts of the world, of methodological contributions made elsewhere. This holds both in a geographical and a disciplinary sense; far too often we see virtually identical techniques re-invented in different disciplines. We need to break down disciplinary as well as international boundaries!

While some chapters in this volume (for instance, chapters 3 and 13) have considered links between ill-health and social deprivation, we see merit in further work along these lines, perhaps linked to the new work on health inequalities that is now beginning to be established. How widely used are small area, census-based measures of deprivation in different European states? Is there scope for comparing associations between ill-health and social deprivation in a set of European cities, for example? Are these associations consistent across Europe, and to what extent are differences due to artefact (definitions and measurement) rather than society and culture? It may be possible to link the GIS expertise represented in this volume and elsewhere with other pan-European research and policy developments (both in the health field and in environmental domains) as these unfold in the years ahead.

We need to broaden our concept of health and our set of possible health indicators to enable us to define surfaces of 'health potential', for example showing access to goods and services and environments that promote quality of life, as well as proximity to health-threatening locations (busy main roads, hazardous waste sites, contaminated water courses, and so on). Certainly in North America there is a burgeoning interest in environmental equity; the extent to which certain social groups are disadvantaged by virtue of their access to environmental 'bads'. In other words, we require maps of what David Harvey (1973) called 'real income'. Where are the 'troughs' on such surfaces, and what can policy-makers do to even out this invisible topography? Implicit in this is a much broader concept of health, seen in terms not merely of mortality but in terms of quality of life and mental well-being. We alluded in our introductory chapter to the need to look not only at residential location as a geographic marker, but at the activity spaces and 'time geographies' (see Löytönen chapter 7) of the individuals whose health experiences we purport to understand.

Extending this theme, what is the scope for embedding a more qualitative dimension into our investigations? Simply recording an individual with chronic illness as a point on a map is, to put it mildly, a crude and simplistic representation of disease. There is a recognition among many health researchers that lay perceptions of health and illness should be given some priority (Popay and Williams, 1994). That said, is there scope for incorporating, for example, taped recordings or oral histories of health experience? Can we not go beyond the GIS workstation as a visual tool and use the power of multimedia in order to incorporate photographic images of what places were like in the past, video footage of pollution events, and transcripts of interviews with those we are studying? Elsewhere, there is a growing interest in the role that GIS might play in an understanding of disabilities, such as visual handicap (Golledge, 1993), and in how the technology might support interfaces other than those based on sight. Here, as elsewhere, there are profitable links to be forged with those working in adjacent fields of the GIS landscape.

Last, those of us interested in GIS and health should not underestimate the need that remains for diffusing into other disciplines, and among other audiences, a geographical perspective. We continue to be surprised at the number of quite senior health professionals who have yet to encounter GIS, or even to consider the merits of the 'geographical imagination' in health research. As we seek to push back the frontiers of scientific knowledge in this important field of human endeavour, we must not neglect the need to seek opportunities to persuade those with pursestrings, and those charged with the responsibilities of monitoring and improving the public health, that geography matters!

References

GOLLEDGE R. G. (1993) Geography and the disabled: a survey with special reference to vision-impaired and blind populations. *Transactions, Institute for British Geographers*, **18**, 63–85.

HARVEY D. (1973) *Social Justice and the City*. Blackwell, Oxford.

POPAY J. and WILLIAMS G. (1994) (eds) *Researching the People's Health*. Routledge, London.

VAN DER VEEN A. A., KUIJPERS-LINDE M. and MEIJER E. N. (1994) DISCET: a system for exchange and presentation of environmental and health data. Pages 1648–1658 in *Proceedings of the 5th European Conference and Exhibition on Geographical Information Systems, EGIS*, Paris, France.

Index

access, 3, 7, 10, 23, 125, 182, 184–5, 187, 206
accessibility, 9, 40, 65, 103, 184
accuracy, 13, 29, 38–9, 65, 81, 98, 168, 192–3, 200–1
activity spaces, 7, 12, 25, 206
address-match, 18, 65, 74
address-matched data, 11, 66–7
address-matched record, 66–7, 69
air pollutants, 5–6, 81, 83, 85–6, 93, 175
air pollution, 5–6, 11, 22, 24, 31, 35–7, 44, 51, 81–3, 85, 87–8, 92–3, 105, 114, 167, 175, 182, 186–7, 205
air quality, 11, 22, 25, 44, 81–5, 92–3
amendment vectors, 102
analysis of disease patterns, 76
area-based data, 30, 40, 49, 64, 185
areal interpolation, 6, 85, 161
authority constraints, 100
autocorrelation, 128

Barnardisation, 39, 200
Bayes adjustment, 39
Bayesian, 12–13, 116, 118, 141, 145, 150
BEAM (Bayesian Ecological Analysis Method), 116, 144
Bayesian smoothing, 6, 12, 39, 139, 144, 181
binary isotonic regression, 52
Bithell's linear risk score test, 52
box plots, 10, 32–3, 44

cancer, 4–6, 8–9, 13–14, 22–3, 25, 39–40, 44, 98, 139–40, 155, 167–76, 181–2, 205
cancer incidence, 4, 14, 42–3, 107, 118, 135, 170–4, 176
cancer maps, 168, 172
cancer registries, 13, 107, 155, 167–70, 172, 174–6, 180–1
cancer statistics, 168–9
capability constraints, 99
cartograms, 8
CD-ROM, 11, 64, 72, 75, 129
census, 6, 9, 14, 24–5, 38–9, 44, 65–6, 74, 84, 107, 113, 119, 127, 180, 184, 191–200
Chernobyl accident, 12, 14, 18, 25, 97, 101, 105–7, 174
cloud plots, 33

cluster alarms, 11, 50–1, 55–7
CEPP (Cluster Evaluation Permutation Procedure), 55
clusters, detection of, 5, 7, 11, 14, 50–1, 53, 55, 57, 69, 76, 162–3
clustering, 5, 7–8, 11–13, 23–4, 50, 52–4, 58–9, 71, 128, 135, 141, 145–8, 150, 162, 176, 179
cluster investigations, 69, 75–6
cluster of activities, 99–101
computing disease rate maps, 72–3
computing Monte Carlo simulations, 11, 73
confirmatory analysis, 56, 63
confirmatory goals, 74
connective tissue, 13, 139–40, 142–4, 147–8, 150
correlation, 162
coupling constraints, 100
Cuzick and Edward's test, 26, 53, 59

database query, 103, 153
death certificate, 127–8, 139, 149
decision support, 4, 120
decision support system, 14, 74–5, 120
demographic structure, 199
density equalised map projection, 54–5
deprivation, 9, 31, 34–5, 38–9, 41, 44, 182, 185, 187, 197–8, 206
deprivation indices, 9, 197–8
descriptive epidemiology, 125, 168
diagnosis of cancer, 167–72
diffusion tube, 83–5, 92
Diggle and Chetwynd's test, 54, 57
disease atlases 6, 55, 76, 132, 149
disease clusters, 5, 7–8, 11, 24–5, 50, 54, 65, 69, 75–6, 179, 185
disease mapping, 18–19, 125, 140, 181–2, 187
disease rates, 19, 31, 35–6, 38, 44, 55–6, 63, 66–7, 69, 71–2, 74–5, 155, 161, 172–4, 179, 181–2, 185
disease surveillance, 11, 55, 74, 179
dispersion modelling, 11, 82, 85–6
domain, 11, 100–1

ecological fallacy, 34, 149, 182
ecological regression analysis, 140, 148, 172
edge effects, 60
electronic atlas, 12, 126, 129, 135–6
empirical Bayes estimation, 6, 12, 139, 161, 181

enabling technology, 17, 19
enumeration districts, 14, 30, 36, 38–40, 44, 116, 184, 191, 199–200
environmental epidemiology, 3, 5–6, 8, 11–12, 14, 19–20, 22–3, 26, 85, 113, 188, 191
environmental hazards, 179, 185, 187–8, 206
environmental monitoring, 115, 122
environmental risk factors, 116
environment and health, 3, 5, 12, 22, 26, 114, 116, 123, 205
epidemiological cancer research, 98, 167–8
evaluation of cluster alarms, 50–1, 55–6
exploratory analysis, 30, 63–4
exploratory and interactive, 64, 74
exploratory data analysis, 10, 30, 155
exploratory spatial analysis, 11, 63–4, 67, 69, 76
exploratory spatial data analysis, 30–1, 38, 44
ESDA (Exploratory Spatial Data Analysis)
exposure, 5–7, 11–14, 17–19, 22, 25–6, 29, 37, 52, 82–5, 92–3, 97–9, 101, 103–7, 115–16, 119–21, 148, 150–1, 167–8, 170–1, 174–6, 181–2, 186–8
exposure assessment, 14, 18–19
extra Poisson, 128, 140–1, 148, 150

false positives, 72
fertilisers, 141, 145–6, 148
fixed effects, 141
flexible spatial filtering, 67
focused tests, 32, 50, 52, 56
food poisoning, 6

GAM, 40
Geary test, 33
'gee whiz' effect, 10, 17, 22–3, 26
general practitioners (GPs), 9, 14, 184, 192–4, 198, 200
general tests, 32
GLMM (Generalised Linear Mixed Models), 140–1, 150
geographical analysis machine, 69
geographical correlation, 181, 185
geographical epidemiology, 5–8, 11–12, 113, 135
geographical pattern, 7, 63, 69, 74, 140–1, 148, 196
geographic scale of analysis, 12, 66, 117, 157, 181, 187
Getis-Ord statistics, 32–3, 43–4
Gibbs sampling, 144, 150
global clustering, 50–4, 56, 59
global space-time clustering, 56, 58–9
global tests, 32, 50–1, 53, 56
grid data, 25, 32, 84, 102

health care planning, 4, 8, 14, 19, 84, 122, 179, 182, 184, 187, 200
health services, 9–10, 122, 135, 179, 182, 184, 193, 197–8
health services research, 9, 29–30, 113, 179, 200
heteroscedasticity, 35–6
histogram, 10, 44
hypothesis testing, 17, 20, 23, 26, 30, 59–60, 182

incidence, 4–8, 13, 34, 36–9, 44, 63–4, 67, 69, 122, 139, 167–8

incubation time, 98
indoor radon exposure, 5, 14, 105, 175
inequalities, health, 3–5, 9–10, 29, 181, 187, 206
infant mortality, 11, 64, 69–72
infant mortality rates, 68–72
INFO-MAP, 8, 40
informal inference, 64
interactive analyses, 8, 10–11, 64

kernel estimates, 129
kernel estimation, 7, 11–12, 116, 128, 135
k nearest neighbours, 53, 57, 162
Knox test, 57–8
kriging, 11, 44, 83–5, 87, 89–90, 92
Krümmel, Germany, 56

latency period, 13, 18, 98, 149, 170, 186
Lawson-Waller local score test, 52
lay perceptions, 4, 206
Legionnaires' disease, 7
limit disclosure, 65–7
linking line dispersion models to the GIS, 86
LISAs (Local Indicators of Spatial Association), 32
LISP-STAT, 8, 40
local Moran plots, 33
local tests, 32
log-linear models, 148, 150
Long Island, New York, 56
Los Alamos, New Mexico, 56
low rates, 4, 55, 59
lung cancer, 4, 14, 105, 107, 116, 118, 126, 132–3, 135, 140, 169, 171–3, 175, 186

magnetic fields, 175
MANET, 40
mapping, 3, 6, 40, 82, 86, 101, 140, 149, 181, 187
maps of air quality, 81, 87, 89–90, 92
maps of significance, 72, 74
markers of exposure, 186–7
MCMC (Markov Chain Monte Carlo), 13, 144, 150–1
median polish, 32–3
migration, 12, 18, 39, 98, 104, 107, 126, 186, 192, 195
mis-specified locations, 60
modelling, health impacts, 187–8
modifiable areal unit problems, 34
monitoring, 6, 11, 19, 75, 81–5, 87–9, 92–3, 105, 115, 122, 139, 206
Monte Carlo simulations, 23, 65, 71–2, 74–5
Moran I index, 128
Moran plots, 32–3
Moran tests, 35
mortality, 4, 6, 11–14, 30, 32, 113, 115–16, 119, 122, 125–9, 132–5, 139–41, 149, 151, 154, 167–8, 171, 176, 198, 206
multicollinearity, 37
multilevel-level modelling, 108
multi-level modelling, 9, 85
multiple myeloma, 13, 139–40, 142–3, 147–8, 150, 169
multiple tests of significance, 55, 72
mutagenic drinking water, 174

need indicators, 14, 193, 197–8
non-Hodgkin's lymphomas, 13, 139, 140, 142–3, 147–8, 150
Norfolk, 193–4, 196–7

outliers, 32–3, 44
ozone, 5–6, 83, 187

patient registers, 7, 14, 191–201
pesticides, 12–13, 120–2, 139, 141, 144, 148, 159–60, 162
phytosanitary products, 149
pocket plots, 32–3
point-based health data, 26, 49, 51, 63, 161
point-in-polygon, 66, 191, 193
point patterns, 7–8, 24, 60
Poisson, 36, 53, 60, 66, 116, 141, 144, 148, 159–60, 162
Poisson regression, 36, 174
Popper, Popperian paradigm, 10, 17, 20–1, 23–4
population, 7–8, 10, 14, 22, 24–5, 29–31, 33–4, 38–41, 44, 51, 53–5, 57–8, 60, 67, 69, 72, 74–5, 82, 85, 90, 92–3, 97–8, 101, 113, 115, 119–22, 125–27, 149, 155, 161, 167–9, 171–2, 174–5, 179–82, 184, 186–7, 191–4, 199–200
population estimates, 14, 192–3, 195–8, 200
population estimation, 14, 192–3, 195–6, 198, 200
postcodes, 7, 14, 25, 38–9, 65, 84, 86, 92, 97, 155, 157, 159, 161, 180, 184, 191, 193–4, 196–7, 199
power studies, 52, 54, 57, 59, 75
practice characteristics, 198–200
privacy and confidentiality, 7, 37–9, 65, 161, 201
protect privacy, 65–7, 125
public health, 8–9, 11–12, 14, 17–19, 23, 29, 63–4, 74–5, 81, 92, 113–15, 155, 179, 181, 187, 191, 193
public health surveillance, 18–19, 74–6, 121

qualitative research, 4, 122, 206

random effects, 141, 150
raster data, 25, 102
reference distributions, 69, 71–2, 74
referral patterns, 179, 182
REGARD, 40
regionalisation, 38, 44
register, data, 97, 105
registers, 191–3, 196, 201, 205
regression, 11, 32, 34, 36–7, 51, 85, 87–9, 91–3, 140, 145, 162, 181, 196
regression model with autocorrelated errors, 35
regular geographic surveillance, 76
relative risk measures, 6, 128
residence, 12, 22, 25, 39, 59, 82, 98, 104, 125–8, 150, 167–8, 170–1, 176, 180, 184, 186, 193
residuals, 30, 33–5, 88
residuals outlier tests, 33
respiratory diseases, 6, 12–13, 25, 29, 35–7, 81, 86, 154–5, 158–60, 182
risk analysis, 76
risk assessment, 12, 113, 119–21, 174, 187
risk estimates, 127, 135, 175
risk factors, 29, 36, 37, 75, 99, 115, 117, 122, 139, 148, 170–1, 182, 185

scatter plot, 32–3
Seascale, England, 56
service areas, 184
significance of clusters, 50, 55, 71
significance of disease rates, 66, 73, 75
significance tests, 23, 49–50, 69
simulated disease patterns, 72
'sliding window', 67
small areas, 7–9, 14, 49, 63, 65, 85, 107, 116, 125–7, 135–6, 149, 161, 181, 187, 197, 205–6
smoothed estimators, 116, 139–40, 144, 148, 150
SMRS (Standardised Mortality Ratios), 116, 118, 127–8, 139, 141–2, 144, 148–9
SpaceStat, 40
space-time interaction, 51, 57, 58–9
space-time scan statistics, 57
Spain, 13, 139–40, 142–4, 149
spatial aggregation methods, 67, 75
spatial analysis, 8, 10–11, 13, 29, 45, 63–5, 72, 75, 97, 125–7, 151, 153, 155–6, 159, 161–3
spatial autocorrelation, 6, 12, 19, 31–3, 35–7, 51, 75, 140, 150, 196
spatial autocovariance, 33
spatial concentration, 31–2
spatial covariation, 31–3
spatial data encoding, 65
spatial dimension, 25, 101–2, 108
spatial epidemiology, 5, 29–30, 49
spatial filter, 11, 67–71, 74
spatial heterogeneity, 10, 13
spatial regression, 35, 49, 51
spatial regression model with spatially lagged dependent variable, 36
spatial regression model with spatially lagged independent variable, 37
spatial scales, 29, 66–7, 69, 71, 104, 120–2, 135, 156
spatial scan statistics, 23, 55–7, 66
spatial statistics, 10–13, 18, 22–4, 26, 29–30, 40, 162–3
spatiotemporal model, 105
spatio-temporal patterns, 103
spatio-temporal processes, 102–4, 107–8
S-Plus, 8, 40
spurious correlation, 35
statistical methods, 7, 11, 23–4, 36–7, 49, 51–2, 64, 85, 87, 93, 107, 181
statistical significance of localised rates, 64
stomach cancer, 126, 132, 134–5, 171, 174
Stone's Poisson maximum test, 52
strong inference, 20–2
study populations, identification of, 18–19, 29
Suffolk, 193,4, 198–200
surveillance, 14, 18–19, 125, 179–80, 185, 188
surveillance of clusters, 55, 57

targeting of interventions, 18–19
temporal dimension, 59, 101–3, 108, 153
tests for spatial randomness, 49, 52, 59
thematic maps, thematic mapping, 17, 22–4, 49, 122
TIGER, 65–6
time geography, 12, 97–103, 107–8, 206
time lag, 98
time-space budgets, 101

time-space coordinates, 99, 101
time-space history, 97
time-space lag, 97–8
time-space models, 102
time-space prism, 101, 104, 107–8
Townsend index, 34, 38–9
Townsend scores, 44
transport, 9, 18, 40, 81, 113, 184, 187
trend surface, 31, 33
trend surface models, 33
trend surface model with autocorrelated errors, 33
triad framework, 103

uncertain, uncertainty, uncertain locations, 12, 25–6, 59–60, 159, 176, 180–1, 185
undercoverage, 170, 192

unemployment, 14
unemployment rates, 153, 196–8
updatable health needs indicators, 200
user interface, 129–30, 153, 163

variable spatial filters, 73
variogram clouds, 32
vector model, 102–3
vehicle emissions, 5, 11, 81, 82–4, 88, 92–3
visualisation, 3–4, 8, 12–14, 22, 55, 66–7, 74, 103, 122, 125, 153, 155–7, 159, 162–3

ward, 14, 30, 38–9, 180, 184, 192–7, 199
Whittemore's test, 53, 59
whole map, 10, 37, 55, 161
'whole map' statistics, 10, 32